Fire Alarm Systems

Georgia Institute of Technology – Open to the Sun

The Georgia Tech house is all about sunlight, using it to transform and open up its living space. It was constructed using translucent walls, made of two sheets of polycarbonate that enclose an aerogel filler. Aerogel, sometimes referred to as "solid smoke," is one of the lightest solids known. It is an excellent insulator and is translucent, allowing filtered light to enter the home.

26405-14

Trainees with successful module completions may be eligible for credentialing through NCCER's National Registry. To learn more, go to **www.nccer.org** or contact us at **1.888.622.3720.** Our website has information on the latest product releases and training, as well as online versions of our *Cornerstone* magazine and Pearson's product catalog.

Your feedback is welcome. You may email your comments to **curriculum@nccer.org,** send general comments and inquiries to **info@nccer.org,** or fill in the User Update form at the back of this module.

This information is general in nature and intended for training purposes only. Actual performance of activities described in this manual requires compliance with all applicable operating, service, maintenance, and safety procedures under the direction of qualified personnel. References in this manual to patented or proprietary devices do not constitute a recommendation of their use.

Objectives

When you have completed this module, you will be able to do the following:

1. Define the unique terminology associated with fire alarm systems.
2. Describe the relationship between fire alarm systems and life safety.
3. Explain the role that various codes and standards play in both commercial and residential fire alarm applications.
4. Describe the characteristics and functions of various fire alarm system components.
5. Identify the different types of circuitry that connect fire alarm system components.
6. Describe the theory behind conventional, addressable, and analog fire alarm systems and explain how these systems function.

Performance Task

Under the supervision of the instructor, you should be able to do the following:

1. Connect selected fire alarm system(s).

Trade Terms

The trade terms for this module are extensive. They appear in boldface (blue) type throughout the text, with definitions provided in the Glossary at the end of this module.

Required Trainee Materials

1. Pencil and paper
2. Appropriate personal protective equipment
3. Copy of the latest edition of the *National Electrical Code*®

Note:
NFPA 70®, *National Electrical Code*®, and *NEC*® are registered trademarks of the National Fire Protection Association, Inc., Quincy, MA 02269. All *National Electrical Code*® and *NEC*® references in this module refer to the 2014 edition of the *National Electrical Code*®.

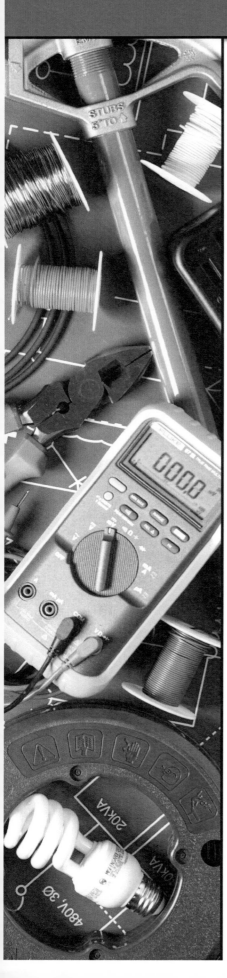

Contents ───────────────────────

Topics to be presented in this module include:

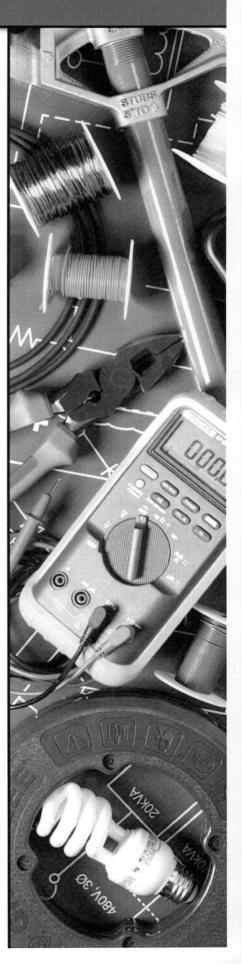

Contents (*continued*) ———————————

Figures and Tables ———————————

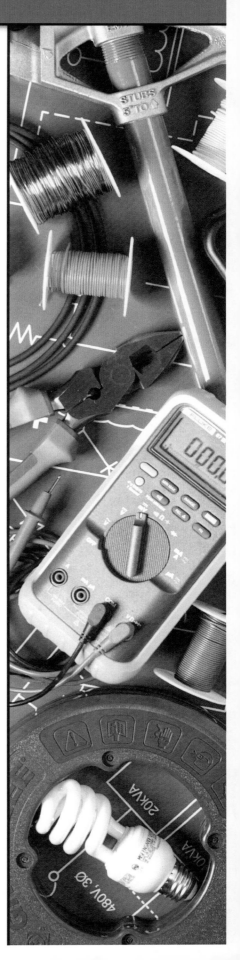

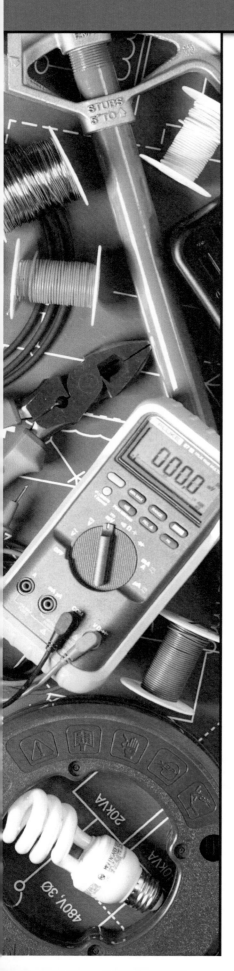

Figures and Tables (*continued*) ───────────

1.0.0 INTRODUCTION

Fire alarm systems can make the difference between life and death. In a fire emergency, a quick and accurate response by a fire alarm system can reduce the possibility of deaths and injuries dramatically. However, practical field experience is necessary to master this trade. Remember to focus on doing the job properly at all times. Failure to do so could have tragic consequences.

Always seek out and abide by any National Fire Protection Association (NFPA), federal, state, and local codes that apply to each and every fire alarm system installation.

Different types of buildings and applications will have different fire safety goals. Therefore, each building will have a different fire alarm system to meet those specific goals. Most automatic fire alarm systems are configured to provide fire protection for life safety, property protection, or mission protection.

Life safety fire protection is concerned with protecting and preserving human life. In a fire alarm system, life safety can be defined as providing warning of a fire situation. This warning occurs early enough to allow notification of building occupants, ensuring that there is sufficient time for their safe evacuation.

Fire alarm systems are generally taken for granted. People go about their everyday business with little or no thought of the safety devices that monitor for abnormal conditions. After all, no one expects a fire to happen. When it does, every component of the life safety fire alarm system must be working properly in order to minimize the threat to life.

2.0.0 CODES AND STANDARDS

Fire alarm system equipment and installation are regulated and controlled by various national, state, and local codes. Industry standards have also been developed as a means of establishing a common level of competency. These codes and standards are set by a variety of different associations, agencies, and laboratories that consist of fire alarm professionals across the country. Depending on the application and installation, you will need to follow the standards established by one or more of these organizations.

In the United States, local and state jurisdictions will select a model code and supporting documents that detail applicable standards. If necessary, they will amend the codes and standards as they see fit. Some amendments may be substantial changes to the code, and in certain situations, the jurisdictions will author their own sets of codes. Once a jurisdiction adopts a set of codes and standards, it becomes an enforceable legal document within that jurisdiction.

The following list details some of the many national organizations responsible for setting the industry standards:

- *Underwriters Laboratories (UL)* – Establishes standards for fire equipment and systems.
- *The National Fire Protection Association (NFPA)* – Establishes standards for fire systems. The association publishes the National Fire Alarm Code® (NFPA 72®), the *Life Safety Code® (NFPA 101®)*, and the *Uniform Fire Code™ (NFPA 1)*. These codes are the primary reference documents used by fire alarm system professionals.
- *Factory Mutual (FM)* – Establishes standards for fire systems.

On Site

Installing and Servicing Fire Alarm Systems

In certain jurisdictions, the person who installs and/or services fire alarm systems must be a licensed fire alarm specialist.

On Site

Property and Mission Protection

The design goal of most fire alarm systems is life safety. However, some systems are designed with a secondary purpose of protecting either property or the activities (mission) within a building. The goal of both property and mission protection is the early detection of a fire so that firefighting efforts can begin while the fire is still small and manageable. Fire alarm systems with a secondary goal of property protection are commonly used in museums, libraries, storage facilities, and historic buildings in order to minimize damage to the buildings or their contents. Systems with a secondary goal of mission protection are commonly used where it is essential to avoid business interruptions, such as in hospitals, financial businesses, security control rooms, and telecommunication centers.

- *National Electrical Manufacturers' Association (NEMA)* – Establishes standards for equipment.
- *The Federal Bank Protection Act* – Establishes fire alarm equipment and system standards for banks.
- *The Defense Intelligence Agency (DIAM-50-3)* – Establishes standards for military and intelligence installations.
- *The National Institute for Certification in Engineering Technologies (NICET)* – Establishes standardized testing of fire alarm designers and installers.

2.1.0 The National Fire Protection Association

The National Fire Protection Association (NFPA) is responsible for setting the national standards and codes for the fire alarm industry. Consisting of fire alarm representatives from all areas of business and fire protection services, the NFPA responds to the ever-changing needs of society by using a democratic process to form consensus standards that are acceptable to all members. The NFPA reviews equipment and system performance criteria and input from experienced industry professionals to set an acceptable level of protection for both life and property.

2.1.1 NFPA Codes

The following four widely adopted national codes are specified by the NFPA:

- *National Electrical Code® (NFPA 70®)* – The *National Electrical Code® (NEC®)* covers all of the necessary requirements for all electrical work performed in a building. The Fire Protective Signaling Systems portion of the code *(NEC Article 760)* details the specific requirements for wiring and equipment installation for fire protection signaling systems. Specifications include installation methods, connection types, circuit identification, and wire types (including gauges and insulation). The *NEC®* places restrictions on the number and types of circuit combinations that can be installed in the same enclosure.
- *National Fire Alarm Code® (NFPA 72®)* – The recommended requirements for installation of fire alarm systems and equipment in residential and commercial facilities are covered in this code. Included are requirements for installation of initiating devices (sensors) and notification appliances (visual or audible). Inspection, testing, and maintenance requirements for fire alarm systems and equipment are also covered.

On Site

NFPA Codes

The three NFPA code books required for installing alarm systems are shown here. Nearly every requirement of *NFPA 101®, Life Safety Code®*, has resulted from the analysis of past fires in which human lives have been lost. The three code books shown were current editions when this module was published. Code books are revised and changed periodically, typically every three or four years. For this reason, you should always make sure that you are using the current edition of any code book.

26405-14_SA01.EPS

- *Life Safety Code® (NFPA 101®)* – This document is focused on the preservation and protection of human life, as opposed to property. Life safety requirements are detailed for both new construction and existing structures. Specifically, necessary protection for unique building features and construction are detailed. In addition, chapters are organized to explain when, where, and for what applications fire alarm systems are required, the necessary means of initiation and occupant notification, and the means by which to notify the fire department. This code also details any equipment exceptions to these requirements.
- *Uniform Fire Code™ (NFPA 1)* – This code was established to help fire authorities continually develop safeguards against fire hazards. A chapter of this code is dedicated to fire protection systems. Information and requirements for testing, operation, installation, and periodic preventive maintenance of fire alarm systems are included in this portion of the code.

2.1.2 NFPA Standards

The NFPA also publishes specific standards that are used by fire alarm system professionals. These include:

- *NFPA 75, Standard for Protection of Electronic Computer/Data Processing Equipment*
- *NFPA 80, Standard for Fire Doors and Other Opening Protectives*
- *NFPA 90A, Standard for the Installation of Air Conditioning and Ventilating Systems*
- *NFPA 90B, Standard for the Installation of Warm Air Heating and Air Conditioning Systems*
- *NFPA 92A, Standard for Smoke-Control Systems Utilizing Barriers and Pressure Differences*

3.0.0 FIRE ALARM SYSTEMS OVERVIEW

Fire alarm systems are primarily designed to detect and warn of abnormal conditions, alert the appropriate authorities, and operate the necessary facility safety devices to minimize fire danger. Through a variety of manual and automatic system devices, a fire alarm system links the sensing of a fire condition with people inside and outside the building. The system communicates to fire professionals that action needs to be taken.

Although specific codes and standards must always be followed when designing and installing a fire alarm system, many different types of systems that employ different types of technology can be used. Most of these types of fire alarm systems are categorized by the means of

communication between the detectors and the fire alarm control panel (FACP). Three major types of fire alarm systems are:

- Conventional hardwired
- Multiplex
- Addressable intelligent

3.1.0 Conventional Hardwired Systems

A hardwired system using conventional initiation devices (heat and smoke detectors and pull stations) and notification devices (appliances) (bells, horns, or lights) is the simplest of all fire alarm systems. Large buildings or areas that are being protected are usually divided into zones to identify the specific area where a fire is detected. A conventional hardwired system is limited to zone detection only, with no means of identifying the specific detector that initiated the alarm. A typical hardwired system might look like *Figure 1* with either two- or four-wire initiating or notification device circuits. In two-wire circuits, power for the devices is superimposed on the alarm circuits. In four-wire circuits, the operating power is supplied to the devices separately from the signal or alarm circuits. In either two- or four-wire systems, end-of-line (EOL) devices are used by the FACP to monitor circuit integrity.

3.2.0 Multiplex Systems

Multiplex systems are similar to hardwired systems in that they rely on zones for fire detection. The difference, however, is that multiplexing allows multiple signals from several sources to be sent and received over a single communication line. Each signal can be uniquely identified. This results in reduced control equipment, less wiring infrastructure, and a distributed power supply. *Figure 2* shows a simplified example of a multiplex system.

3.3.0 Addressable and Analog Addressable Systems

Two different versions of addressable intelligent systems are available. They are addressable and analog addressable systems. Analog addressable systems are more sophisticated than addressable systems.

3.3.1 Addressable Systems

Addressable systems use advanced technology and detection equipment for discrete identification of alarm signals at the detector level. An addressable system can pinpoint an alarm

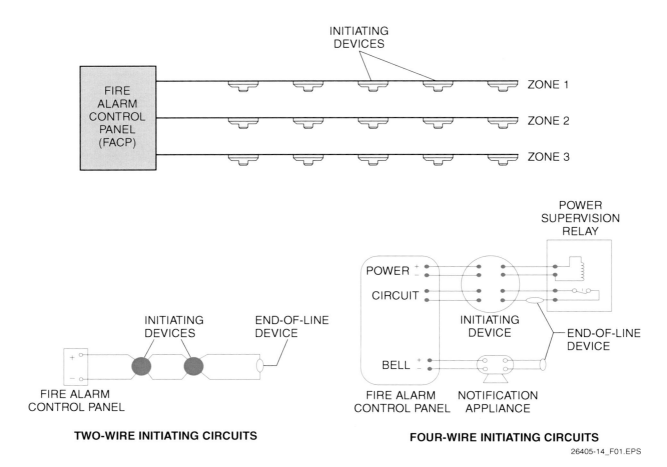

Figure 1 Typical conventional hardwired system.

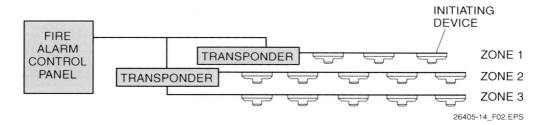

Figure 2 Typical multiplex system.

location to the precise physical location of the initiating detector. The basic idea of an addressable system is to provide identification or control of individual initiation, control, or notification devices on a common circuit. Each component on the signaling line circuit (SLC) has an identification number or address. The addresses are usually assigned using switches or other similar means. *Figure 3* is a simplified representation of an addressable or analog addressable fire alarm system.

The fire alarm control panel (FACP) constantly polls each device using a signaling line circuit (SLC). The response from the device being polled verifies that the wiring path (pathway) is intact (wiring supervision) and that the device is in place and operational. Most addressable systems use at least three states to describe the status of the device: normal, trouble, and alarm. Smoke detection devices make the decision internally regarding their alarm state just like conventional smoke detectors. Output devices, like relays, are also checked for their presence and in some cases for their output status. Notification output modules also supervise the wiring to the horns, strobes, and other devices, as well as the availability of

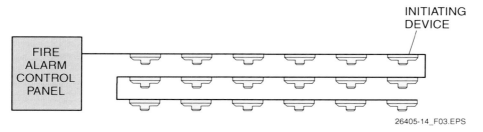

Figure 3 Typical addressable or analog addressable system.

the power needed to run the devices in case of an alarm. When the FACP polls each device, it also compares the information from the device to the system program. For example, if the program indicates device 12 should be a contact transmitter but the device reports that it is a relay, a problem exists that must be corrected. Addressable fire alarm systems have been made with two, three, and four conductors. Generally, systems with more conductors can handle more addressable devices. Some systems may also contain multiple SLCs. These are comparable to multiple zones in a conventional hardwired system.

3.3.2 Analog Addressable Systems

Analog systems take the addressable system capabilities much further and change the way the information is processed. When a device is polled, it returns much more information than a device in a standard addressable system. For example, instead of a smoke detector transmitting that it is in alarm status, the device actually transmits the level of smoke or contamination present to the fire alarm control panel. The control panel then compares the information to the levels detected in previous polls. A slow change in levels (over days, weeks, or months) indicates that a device is dirty or malfunctioning. A rapid change, however, indicates a fire condition. Most systems have the capability to compensate for the dirt buildup in the detectors. The system will adjust the detector sensitivity to the desired range. Once the dirt buildup exceeds the compensation range, the system reports a trouble condition. The system can also administer self-checks on the detectors to test their ability to respond to smoke. If the airflow around a device is too great to allow proper detection, some systems will generate a trouble report.

The information in some systems is transmitted and received in a totally digital format. Others transmit the polling information digitally but receive the responses in an analog current-level format.

The panel, not the device, performs the actual determination of the alarm state. In many systems,

the light-emitting diode (LED) on the detector is turned on by the panel and not by the detector. This ability to make decisions at the panel also allows the detector sensitivity to be adjusted at the panel. For instance, an increase in the ambient temperature can cause a smoke detector to become more sensitive, and the alarm level sensitivity at the panel can be adjusted to compensate. Sensitivities can even be adjusted based on the time of day or day of the week. Other detection devices can also be programmed to adjust their own sensitivity.

The ability of an analog addressable system to process more information than the three elementary alarm states found in simpler systems allows the analog addressable system to provide pre-alarm signals and other information. In many devices, five or more different signals can be received.

Most analog addressable systems operate on a two-conductor circuit. Most systems limit the number of devices to about one hundred. Because of the high data rates on these signaling line circuits, capacitance also limits the conductor lengths. Always follow the manufacturer's installation instructions to ensure proper operation of the system.

3.3.3 Signaling Line Circuits (SLCs)

In addressable or analog addressable systems, there are two basic wiring or circuit types (classes). Class B is the most common and has six different styles. Class A, with four styles, provides additional reliability but is not normally required and is generally more expensive to install. Various codes will address what circuit types are required. The performance requirements for each of the styles of SLC circuits are detailed in *NFPA 72*®.

- Class B circuits – A Class B signaling line circuit for an addressable system essentially requires that two conductors reach each device on the circuit by any means as long as the wire type and physical installation rules are followed. It does not require wiring to pass in and out of each device in a series arrangement. With a circuit capable of 100 devices, it is permissible to go in 100 different directions from the panel. Supervision occurs because the fire alarm control panel polls and receives information from each device. The route taken is not important. A break in the wiring will result in the loss of communication with one or more devices.

- Class A circuits – A Class A signaling line circuit for an addressable system requires that the conductors loop into and out of each device. At the last device, the signaling line circuit is returned to the control panel by a different route. The control panel normally communicates with the devices via the outbound circuit but has the ability to communicate to the back side of a break through the return circuit. The panel detects the fact that a complete loop no longer exists and shifts the panel into Class A mode. All devices remain in operation. Some systems even have the ability to identify which conductor has broken and how many devices are on each side of the break.
- *Hybrid circuits* – A hybrid system may consist of a Class A main trunk with Class B spur circuits in each area. A good example would be a Class A circuit leaving the control panel, entering a junction box on each floor of a multistory building, and then returning to the control panel by a different route. The signaling line circuits on the floors are wired as Class B from the junction box. This provides good system reliability and keeps costs down.

4.0.0 FIRE ALARM SYSTEM EQUIPMENT

The equipment used in fire alarm systems is generally held to higher standards than typical electrical equipment. The main components of a fire alarm system include:

- Alarm initiating devices
- Control panels
- Primary (main) and secondary (standby) power supplies
- Notification appliances
- Communications and monitoring

5.0.0 FIRE ALARM INITIATING DEVICES

Fire alarm systems use initiating devices to report a fire and provide supervisory or trouble reports. Some of the initiating devices used to trigger a fire alarm are designed to sense the signs of fire automatically (automatic sensors). Some report level-of-fire conditions only. Others rely on people to see the signs of fire and then activate a manual

device. Automatic sensors (detectors) are available that sense smoke, heat, and flame. Manual initiating devices are usually some form of pull station.

Figure 4 is a graphic representation of the application of various detectors for each stage of a fire. However, detection may not occur at any specific point in time within each stage, nor will the indicated sensors always provide detection in the stages represented. For example, alcohol fires can produce flame followed by heat while producing little or no visible particles.

Besides triggering an alarm, some fire alarm systems use the input from one or more detectors or pull stations to trigger fire suppression equipment such as a carbon dioxide (CO_2), dry chemical, or water deluge system.

It is mandatory that devices used for fire detection be listed for the purpose for which they will be used. Common UL listings for fire detection devices are:

- *UL 38* – Manually actuated signaling boxes
- *UL 217* – Single- and multiple-station smoke detectors
- *UL 268* – System-type smoke detectors
- *UL 268A* – Duct smoke detectors
- *UL 521* – Heat detectors

5.1.0 Conventional versus Addressable Commercial Detectors

Figure 5 shows some typical commercial detectors. Commercial detectors differ from stand-alone residential detectors in that a number of commercial detectors are usually wired in parallel and connected to a fire alarm control panel. In a typical conventional commercial fire alarm installation, a number of separate zones with multiple automatic sensors are used to partition the installation into fire zones.

As described previously, newer commercial fire systems use addressable automatic detectors, pull stations, and notification appliances to supply a coded identification signal. The systems also provide individual supervisory or trouble signals to the FACP, as well as any fire alarm signal, when periodically polled by the FACP. This provides the fire control system with specific device location information to pinpoint the fire, along with detector, pull station, or notification appliance status information. In many new commercial fire systems, analog addressable detectors (non-automatic) are used. Instead of sending a fire alarm signal when polled, these devices communicate information about the fire condition (level of smoke, temperature, etc.) in addition to their

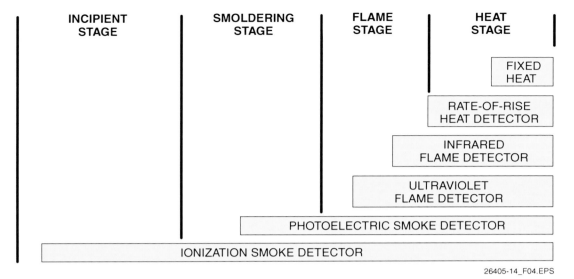

Figure 4   Detection versus stages of fires.

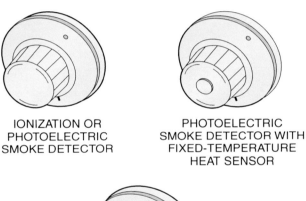

IONIZATION OR
PHOTOELECTRIC
SMOKE DETECTOR

PHOTOELECTRIC
SMOKE DETECTOR WITH
FIXED-TEMPERATURE
HEAT SENSOR

RATE-OF-RISE HEAT DETECTOR WITH
FIXED-TEMPERATURE HEAT SENSOR

26405-14_F05.EPS

Figure 5   Typical commercial automatic sensors.

identification, supervisory, or trouble signals. In the case of these detectors, the control panel's internal programming analyzes the fire condition data from one or more sensors, using recent and historical information from the sensors, to determine if a fire alarm should be issued.

<table>
<tr><td>NOTE</td><td>The descriptions of operation discussed in the sections covering automatic detectors, pull stations, and notification appliances are primarily for conventional devices.</td></tr>
</table>

5.2.0 Automatic Detectors

Automatic detectors can be divided into the following types:

- *Line detectors* – Detection is continuous along the entire length of the detector in these detection devices. Typical examples may include certain older pneumatic rate-of-rise tubing detectors, projected beam smoke detectors, and heat-sensitive cable.
- *Spot detectors* – These devices have a detecting element that is concentrated at a particular location. Typical examples include bimetallic detectors, fusible alloy detectors, certain rate-of-rise detectors, certain smoke detectors, and thermoelectric detectors.
- Air sampling detectors – These consist of piping or tubing distributed from the detector unit to the area(s) to be protected. An air pump draws air from the protected area back to the detector through air sampling ports and piping or tubing. At the detector, the air is analyzed for fire products.
- *Addressable or analog addressable detectors* – As mentioned previously, these detectors provide alarms and individual point identification, along with supervisory/trouble information, to the control panel. In certain analog addressable detectors, adjustable sensitivity of the alarm signal can be provided. In other analog addressable detectors, an alarm signal is not generated. Instead, only a level of detection signal is fed to the control panel. Alarm sensitivity can be adjusted, and the level of detection can be analyzed from the panel (based on historical data) to reduce the likelihood of false alarms in construction areas or areas of high humidity

on a temporary or permanent basis. Because all detectors are continually polled, a T-tap splice is permitted with some signaling line circuit styles. T-taps are not permitted with all styles.

5.3.0 Heat Detectors

There are two major types of heat detectors in general use. One is a rate-of-rise detector that senses a 15°-per-minute increase in room/area temperature. The other is one of several versions of fixed-temperature detectors that activate if the room/area exceeds the rating of the sensor. Generally, rate-of-rise sensors are combined with fixed-temperature sensors in a combination heat detector. Heat detectors are generally used in areas where property protection is the only concern or where smoke detectors would be inappropriate.

5.3.1 Fixed-Temperature Heat Detectors

Fixed-temperature heat detectors activate when the temperature exceeds a preset level. Detectors are made to activate at different levels. The most commonly used temperature settings are 135°F, 190°F, and 200°F. The three types of fixed-temperature heat detectors are:

- *Fusible link* – The fusible link detector (*Figure 6*) consists of a plastic base containing a switch mechanism, wiring terminals, and a three-disc heat collector. Two sections of the heat collector are soldered together with an alloy that will cause the lower disc to drop away when the rated temperature is reached. This moves a plunger that shorts across the wiring contacts, causing a constant alarm signal. After the detector is activated, a new heat collector must be installed to reset the detector to an operating condition.

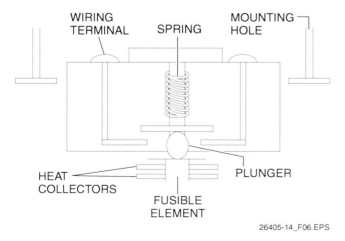

Figure 6 Fusible link detector.

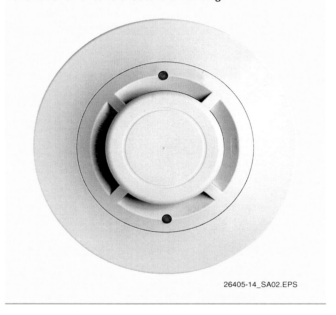
- *Quick metal* – The operation of the quick metal detector (*Figure 7*) is very simple. When the surrounding air reaches the prescribed temperature (usually 135°F or 190°F), the quick metal begins to soften and give way. This allows spring pressure within the device to push the top portion of the thermal element out of the way, causing the alarm contacts to close. After the detector has been activated, either the detector or the heat collector must be replaced to reset the detector to an operating condition.
- *Bimetallic* – In a bimetallic detector (*Figure 8*), two metals with different rates of thermal expansion are bonded together. Heat causes the two metals to expand at different rates, which causes the bonded strip to bend. This action closes a normally open circuit, which signals an alarm. Detectors with bimetallic elements automatically self-restore when the temperature returns to normal.

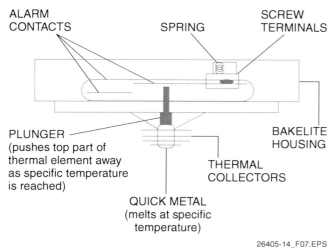

ALARM CONTACTS SPRING SCREW TERMINALS

PLUNGER
(pushes top part of
thermal element away
as specific temperature
is reached)

BAKELITE HOUSING

THERMAL COLLECTORS

QUICK METAL
(melts at specific
temperature)

26405-14_F07.EPS

Figure 7 Quick metal detector.

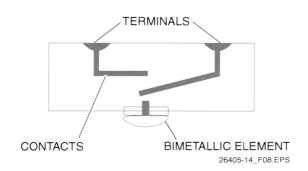

TERMINALS

CONTACTS BIMETALLIC ELEMENT

26405-14_F08.EPS

Figure 8 Bimetallic detector.

5.3.2 Combination Heat Detectors

Combination heat detectors contain two types of heat detectors in a single housing. A fixed-temperature detector reacts to a preset temperature, and a rate-of-rise detector reacts to a rapid change in temperature even if the temperature reached does not exceed the preset level.

- *Rate-of-rise with fusible link detector* – (See *Figure 9*.) Rate-of-rise operation occurs when air in the chamber (1) expands more rapidly than it can escape from the vent (2). The increasing pressure moves the diaphragm (3) and causes the alarm contacts (4 and 5) to close, which results in an alarm signal. This portion of the detector automatically resets when the temperature stabilizes. A fixed-temperature trip occurs when the heat causes the fusible alloy (6) to melt, which releases the spring (7). The spring depresses the diaphragm, which closes the alarm contacts (4 and 5). If the fusible alloy is melted, the center section of the detector must be replaced.

- *Rate-of-rise with bimetallic detector* – (See *Figure 10*.) Rate-of-rise operation occurs when the air in the air chamber (1) expands more rapidly than it can escape from the valve (2). The increasing pressure moves the diaphragm (3) and causes the alarm contact (4) to close. A fixed-temperature trip occurs when the bimetallic element (5) is heated, which causes it to bend and force the spring-loaded contact (6) to mate with the fixed contact (7).

5.3.3 Heat Detector Ratings

As with all automatic detectors, heat sensors are also rated for a listed spacing as well as a temperature rating (*Table 1*). The listed spacing is typically 50' × 50'. The applicable rules and formulas for proper spacing (from *NFPA 72®* and other applicable standards) are then applied to the listed spacing. Caution is advised because it is difficult or impossible to distinguish the difference in the listed spacing for two different types of heat sensors based on their appearance.

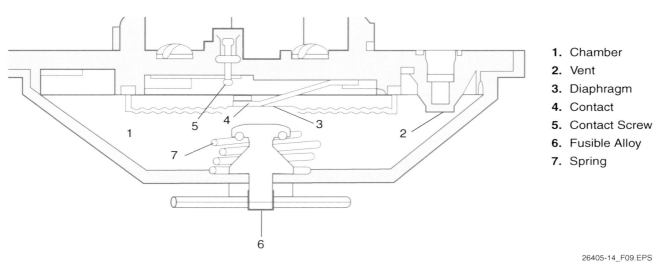

1. Chamber
2. Vent
3. Diaphragm
4. Contact
5. Contact Screw
6. Fusible Alloy
7. Spring

26405-14_F09.EPS

Figure 9 Rate-of-rise with fusible link detector.

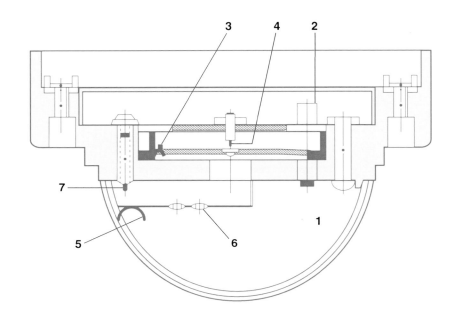

1. Air Chamber
2. Breather Valve
3. Diaphragm Assembly
4. Rate-of-Rise Contact
5. Bimetal Fixed-Temperature Element
6. Contact Spring
7. Fixed-Temperature Contact

26405-14_F10.EPS

Figure 10   Rate-of-rise with bimetallic detector.

Table 1   Heat Detector Temperature Ratings

Temperature Classification	Temperature Rating Range	Maximum Ceiling Temperature	Color Code
Low	100°F–134°F	80°F	No color
Ordinary	135°F–174°F	100°F	No color
Intermediate	175°F–249°F	150°F	White
High	250°F–324°F	225°F	Blue
Extra high	325°F–399°F	300°F	Red
Very extra high	400°F–499°F	375°F	Green
Ultra high	500°F–575°F	475°F	Orange

The maximum ceiling temperature must be 20° or more below the detector rated temperature. The difference between the rated temperature and the maximum ambient temperature for the space should be as small as possible to minimize response time.

Heat sensors are not considered life safety devices. They should be used in areas that are unoccupied or in areas that are environmentally unsuitable for the application of smoke detectors. To every extent possible, heat sensors should be limited to use as property protection devices.

5.4.0 Smoke Detectors

There are two basic types of smoke detectors: photoelectric detectors (*Figure 11*), which sense the presence of smoke, and ionization detectors (*Figure 12*), which sense the presence of combustible gases. For residential use, both types are available combined in one detector housing for maximum coverage. For commercial use, either type is also available combined with a heat detector. As mentioned previously, the newer commercial fire systems use smoke detectors that are analog addressable units, which do not signal an alarm. Instead, they return a signal that represents the level of detection to the FACP for further analysis.

5.4.1 Ionization Detectors

An ionization detector (*Figure 13*) uses the change in the electrical conductivity of air to detect smoke. An alarm is indicated when the amount of smoke in the detector rises above a certain level. The detector has a very small amount of radioactive material in the sensing chambers. As shown in *Figure 14A*, the radioactive material ionizes the air in the measuring and reference chambers, allowing the air to conduct current through the space between two charged electrodes. When smoke particles enter the measuring chamber, they prevent the air from conducting as much current (*Figure 14B*). The detector activates an alarm when the conductivity decreases to a set level. The detector compares the current drop in the main chamber against the drop in the reference chamber. This allows it to avoid alarming when the current drops due to surges, radio frequency interference (RFI), or other factors.

5.4.2 Photoelectric Smoke Detectors

There are two basic types of photoelectric smoke detectors: light-scattering detectors and beam detectors. Light-scattering detectors use the

26405-14_F11.EPS

Figure 11   Typical photoelectric smoke detector.

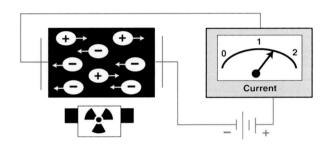

(A) CLEAN AIR

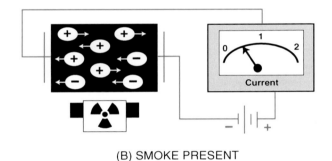

(B) SMOKE PRESENT

26405-14_F14.EPS

Figure 14   Ionization action.

26405-14_F12.EPS

Figure 12   Typical ionization smoke detector.

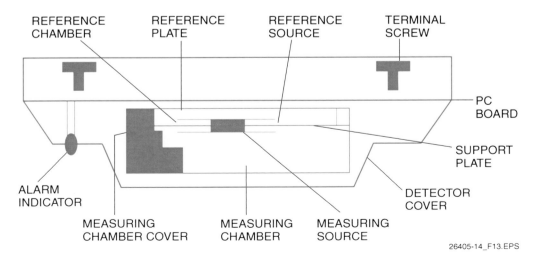

REFERENCE CHAMBER   REFERENCE PLATE   REFERENCE SOURCE   TERMINAL SCREW

PC BOARD

SUPPORT PLATE

DETECTOR COVER

ALARM INDICATOR

MEASURING CHAMBER COVER   MEASURING CHAMBER   MEASURING SOURCE

26405-14_F13.EPS

Figure 13   Ionization detector.

Photoelectric and Ionization Detectors

Photoelectric detectors are often faster than ionization detectors in sensing smoke from slow, smoldering fires. Ionization detectors are often better than photoelectric detectors at sensing fast, flaming fires.

reflective properties of smoke to detect the smoke. The light scattering principle is used for the most common single-housing detectors. Beam detectors rely upon smoke to block enough light to cause an alarm. Early smoke detectors operated on the obscuration principle. Today, photoelectric smoke detectors are used primarily to sense smoke in large open areas with high ceilings. In some cases, mirrors are used to direct the beam in a desired path; however, the mirrors reduce the overall range of the detector. Most commercial photoelectric smoke detectors require an external power source.

Light-scattering detectors are usually spot detectors that contain a light source and a photosensitive device arranged so that light rays do not normally fall on the device (*Figure 15*). When smoke particles enter the light path, the light hits

the particles and scatters, hitting the photosensor, which signals the alarm.

Projected beam smoke detectors that operate on the light obscuration principle (*Figure 16*) consist of a light source, a light beam focusing device, and a photosensitive device. Smoke obscures or blocks part of the light beam, which reduces the amount of light that reaches the photosensor, signaling an alarm. The prime use of the light obscuration principle is with projected beam-type smoke detectors that are employed in the protection of large, open high-bay warehouses, high-ceiling churches, and other large areas. A version of a beam detector was used in older style duct detectors.

After a number of fires where smoke spread through building duct systems and contributed to numerous deaths, building codes began to require the installation of duct detectors. Two versions of duct detectors are shown in *Figure 17*. The projected beam detector is an older style of duct detector that may be encountered in existing installations.

Duct detectors enable a system to control the spread of smoke within a building by turning off the HVAC system, operating exhaust fans, closing doors, or pressurizing smoke compartments in the event of a fire. This prevents smoke, fumes, and fire by-products from circulating through the ductwork. Because the air in the duct is either at rest or moving at high speed, the detector must be able to sense smoke in either situation. A typical duct detector has a listed airflow range within which it will function properly. It is not required to function when the duct fans are stopped.

At least one duct detector is available with a small blower for use in very low airflow applications. Duct detectors must not be used as a substitute for open area protection because smoke

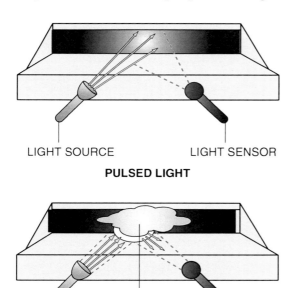

PULSED LIGHT

SCATTERED LIGHT

26405-14_F15.EPS

Figure 15 Light-scattering detector operation.

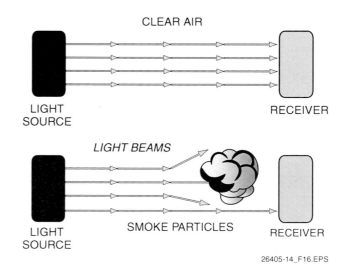

CLEAR AIR

LIGHT SOURCE

RECEIVER

LIGHT BEAMS

LIGHT SOURCE

SMOKE PARTICLES

RECEIVER

26405-14_F16.EPS

Figure 16 Light obscuration principle.

Projected Beam Smoke Detectors

The transmitter and receiver for a projected beam smoke detector are shown here. These units are designed for use in atriums, ballrooms, churches, warehouses, museums, factories, and other large, high-ceiling areas where conventional smoke detectors cannot be easily installed.

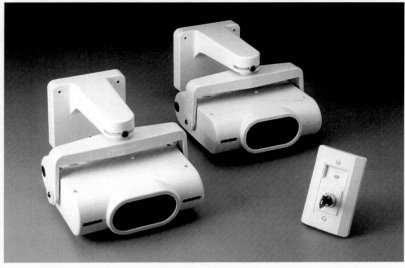

26405-14_SA03.EPS

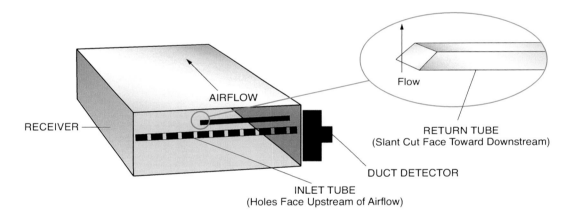

RECEIVER

AIRFLOW

Flow

RETURN TUBE
(Slant Cut Face Toward Downstream)

DUCT DETECTOR

INLET TUBE
(Holes Face Upstream of Airflow)

SAMPLING TUBE STYLE

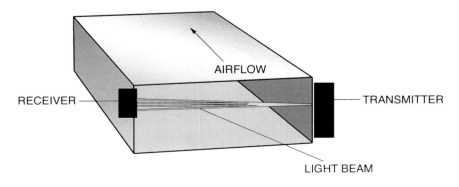

AIRFLOW

RECEIVER

TRANSMITTER

LIGHT BEAM

PROJECTED LIGHT BEAM STYLE

26405-14_F17.EPS

Figure 17   Typical duct detector installations.

Duct Smoke Detectors

This is a typical duct smoke detector with the cover removed. It is designed for HVAC applications. The manufacturer of this unit makes duct detectors based on either photoelectric technology or ionization technology. Up to ten of these units can be networked together. When one detector goes into alarm, all of the other detectors on the network activate in order to control the connected fans, blowers, and dampers. However, only the duct detector that initiated the alarm shows an alarm indication to identify it as the source of the alarm.

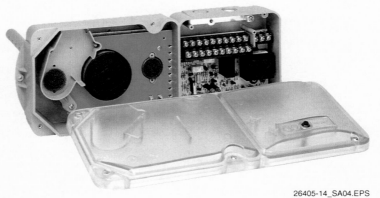

26405-14_SA04.EPS

may not be drawn from open areas of the building when the HVAC system is shut down. Duct detectors can be mounted inside the duct with an access panel or outside the duct with sampling tubes protruding into the duct.

As mentioned previously, the primary function of a duct detector is to turn off the HVAC system to prevent smoke from being circulated. However, *NFPA 90A* and *NFPA 90B* require that duct detectors be tied into a general fire alarm system if the building contains one. If no separate fire alarm system exists, then remote audio/visual indicators, triggered by the duct detectors, must be provided in normally occupied areas of the building. Duct detectors that perform functions other than the shutdown of HVAC equipment must be supplied with backup power.

Cloud chamber smoke detectors (*Figure 18*) use sampling tubes to draw air from several areas (zones). The air is passed through several

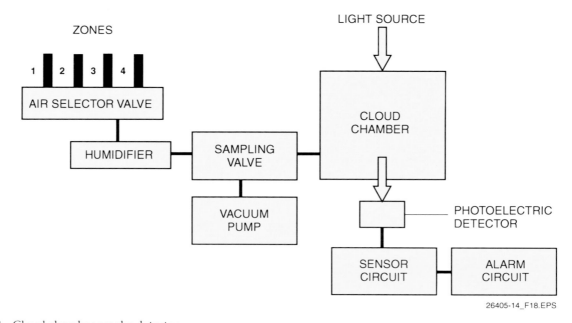

Figure 18 Cloud chamber smoke detector.

chambers where humidity is added and pressure is reduced with a vacuum pump. Reducing the pressure causes water droplets to form around the sub-micron smoke particles and allows them to become visible. A light beam is passed through the droplets and measured by a photoelectric detector. As the number of droplets increases, the light reaching the detector is reduced, which initiates an alarm signal.

5.5.0 Other Types of Detectors

This section describes special-application detectors. These include rate compensation detectors, semiconductor line-type heat detectors, fusible line-type heat detectors, ultraviolet flame detectors, and infrared flame detectors.

5.5.1 Rate Compensation Detectors

A rate compensation detector (*Figure 19*) is a device that responds when the temperature of the surrounding air reaches a predetermined level, regardless of the rate of temperature rise. The detector also responds if the temperature rises quickly over a short period. This type of detector has a tubular casing of metal that expands lengthwise as it is heated. A contact closes as it reaches a certain stage of elongation. These detectors are self-restoring. Rate compensation detectors are more complex than either fixed or rate-of-rise detectors. They combine the principles of both to compensate for thermal lag. When the air temperature is rising rapidly, the unit is designed to respond almost exactly at the point when the air temperature reaches the unit's rated temperature. It does not lag while it absorbs the heat and rises to that temperature. Because of the precision associated with their operation, rate compensation detectors are well suited for use in areas where thermal lag must be minimized.

5.5.2 Semiconductor Line-Type Heat Detectors

A restorable semiconductor line-type detector (*Figure 20*) uses a semiconductor material and a stainless steel capillary tube. The capillary tube contains a coaxial center conductor separated from the tube wall by a temperature-sensitive semiconductor thermistor material. Under normal conditions, a small current (below the alarm threshold) flows. As the temperature rises, the resistance of the semiconductor thermistor decreases. This allows more current to flow and initiates the alarm. When the temperature falls, the current flow decreases below the alarm threshold level, which stops the alarm. The wire is connected to special controls or modules that establish the alarm threshold level and sense the

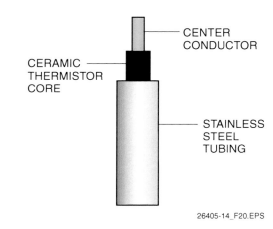

26405-14_F20.EPS

Figure 20 Restorable semiconductor line-type heat detector.

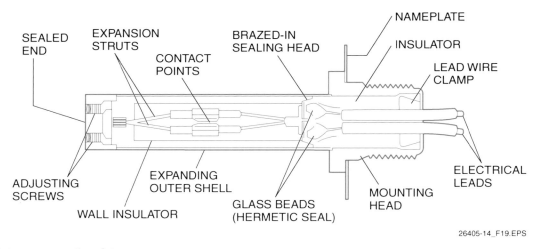

26405-14_F19.EPS

Figure 19 Rate compensation detector.

current flow. Some of these control devices can pinpoint the location in the length of the wire where the temperature change occurs. Line-type heat detectors are commonly used in cable trays, conveyors, electrical switchgear, warehouse rack storage, mines, pipelines, hangars, and other similar applications.

5.5.3 Fusible Line-Type Heat Detector

A non-restorable fusible line-type heat detector (*Figure 21*) uses a pair of steel wires in a normally open circuit. The conductors are held apart by heat-sensitive insulation. The wires, under tension, are enclosed in a braided sheath to make

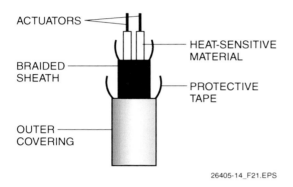

26405-14_F21.EPS

Figure 21   Non-restorable fusible line-type heat detector.

a single cable. When the temperature limit is reached, the insulation melts, the two wires contact, and an alarm is initiated. The melted and fused section of the cable must be replaced following an alarm to restore the system. The wire is available with different melting temperatures for the insulation. The temperature rating should be approximately 20°F above the ambient temperature. Special controls or modules are available that can be connected to the wires to pinpoint the fire location where the wires are shorted together.

5.5.4 Ultraviolet Flame Detectors

An ultraviolet (UV) flame detector (*Figure 22*) uses a solid-state sensing element of silicon carbide, aluminum nitrate, or a gas-filled tube. The UV radiation of a flame causes gas in the element or tube to ionize and become conductive. When sufficient current flow is detected, an alarm is initiated.

5.5.5 Infrared Flame Detectors

An infrared (IR) flame detector (*Figure 23*) consists of a filter and lens system that screens out unwanted radiant-energy wavelengths and focuses the incoming energy on light-sensitive components. These flame detectors can respond to the

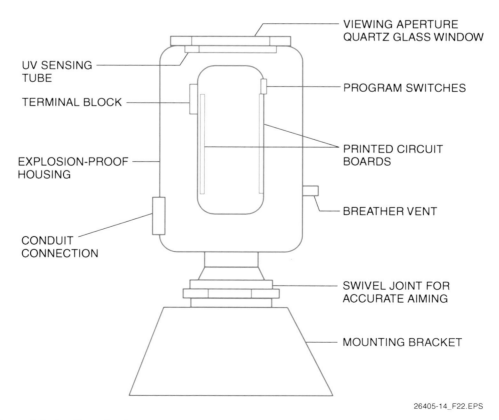

26405-14_F22.EPS

Figure 22   UV flame detector (top view).

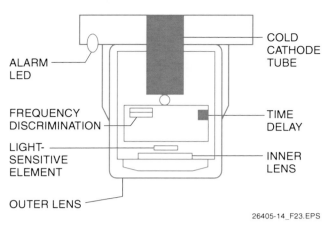

ALARM LED

FREQUENCY DISCRIMINATION

LIGHT-SENSITIVE ELEMENT

OUTER LENS

COLD CATHODE TUBE

TIME DELAY

INNER LENS

26405-14_F23.EPS

Figure 23   IR flame detector (top view).

total IR content of the flame alone or to a combination of IR with flame flicker of a specific frequency. They are used indoors and have filtering systems or solar sensing circuits to minimize unwanted alarms from sunlight.

5.6.0 Manual (Pull Station) Fire Detection Devices

When required by code, manual pull stations are required to be distributed throughout a commercial monitored area so that they are unobstructed, readily accessible, and in the normal path of exit from the area. Examples of pull stations are shown in *Figure 24*.

- *Single-action pull stations* – A single action or motion operates these devices. They are activated by pulling a handle that closes one or more sets of contacts and generates the alarm.
- *Glass-break pull stations* – In these devices, a glass rod, plate, or special element must be broken to

activate the alarm. This is accomplished using a handle or hammer that is an integral part of the station. When the alarm is activated, one or more sets of contacts are closed and an alarm is actuated. Usually, the plate, rod, or element must be replaced to return the unit to service, although some stations will operate without the rod or plate.

- *Double-action pull stations* – Double-action pull stations require the user to lift a cover or open a door before operating the pull station. Two discretely independent actions are required to operate the station and activate the alarm. Using a stopper-type cover that allows the alarm to be tripped after a cover is lifted may turn a single-action pull station into a double-action pull station. According to Underwriters Laboratories, the stopper-type device is listed as an accessory to manual stations and is permissible for use as a double-action pull station device. With certain types of double-action stations, there have been instances when people have confused the sound of a tamper alarm that sounds when the cover is lifted with the sound of fire alarm activation. The sounding of this tamper alarm sometimes causes them to fail to activate the pull station.

SINGLE-ACTION PULL STATION

GLASS-BREAK COVER FOR PULL STATION

DOUBLE-ACTION PULL STATION

26405-14_F24.EPS

Figure 24   Typical pull stations.

- *Key-operated pull stations* – Applications for key-operated pull stations (*Figure 25*) are restricted. Key-operated stations are permitted in certain occupancies where facility staff members may be in the immediate area and where use by other occupants of the area is not desirable. Typical situations would include certain detention and correctional facilities and some health care facilities, particularly those that provide mental health treatment.

5.7.0 Auto-Mechanical Fire Detection Equipment

This section describes various types of auto-mechanical fire detection equipment. It covers wet and dry sprinkler systems and water flow alarms.

5.7.1 Wet Sprinkler Systems

A wet sprinkler system (*Figure 26*) consists of a permanently piped water system under pressure, using heat-actuated sprinklers. When a fire occurs, the sprinkler heads exposed to high heat open and discharge water individually in an attempt to control or extinguish the fire. They are designed to automatically detect and control a fire and protect a structure. Once a sprinkler head is activated, some type of water flow sensor signals a fire alarm. When activated, a sprinkler system may cause water damage. Wet systems should not be used in spaces subject to freezing.

> **WARNING!**
> If the air pressure drops and the system fills with water, the system must be drained. Consult a qualified specialist to restore the system to normal operation.

26405-14_F25.EPS

Figure 25   Key-operated pull station.

Manual Fire Alarm Station Reset

This manual fire alarm is operated by pulling on the pull cover. This engages a latching mechanism that prevents the pull cover from being returned to the closed position. The only way the station cover can be reset to the closed position is by using the appropriate reset key.

26405-14_SA05.EPS

5.7.2 Dry Sprinkler Systems

A dry sprinkler system (*Figure 27*) consists of heat-operated sprinklers that are attached to a piping system containing air under pressure. Normally, air pressure in the pipes holds a water valve closed, which keeps water out of the piping system. When heat activates a sprinkler head, the open sprinkler head causes the air pressure to be released. This allows the water valve to open, and water flows through the pipes and out to the activated sprinkler head. Once a sprinkler head is activated and water starts to flow, a water flow sensor signals a fire alarm. Because the pipes are dry until a fire occurs, these systems may be used in spaces subject to freezing.

NCCER — *Electrical Level Four*   26405-14

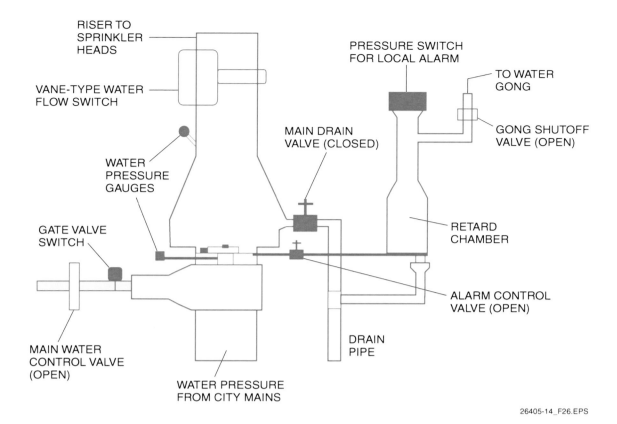

Figure 26 Wet sprinkler system.

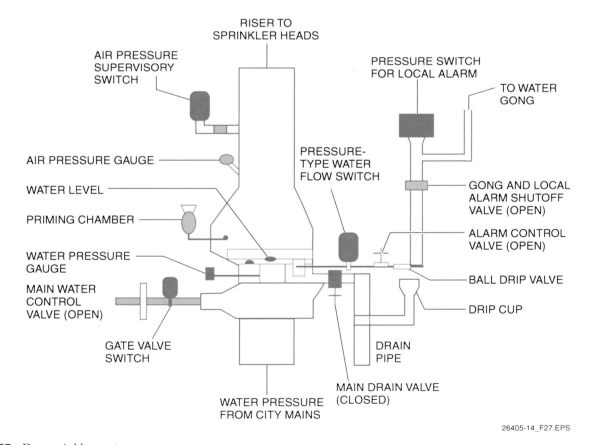

Figure 27 Dry sprinkler system.

5.7.3 Wet or Dry Sprinkler System Water Flow Alarms

When a building sprinkler head is activated by the heat generated by a fire, the sprinkler head allows water to flow. As shown in *Figures 26 and 27*, pressure-type or vane-type water flow switches are installed in the sprinkler system along with local alarm devices. The water flow switches detect the movement of water in the system. Activation of these switches by the movement of water causes an initiation signal to be sent to an FACP that signals a fire alarm.

In addition to devices that signal sprinkler system activation, other devices may be used to monitor the status of the system using an FACP. For instance, the position of a control valve may be monitored so that a supervisory signal is sent whenever the control valve is turned to shut off the water to the sprinkler system. If this valve is turned off, no water can flow through the sprinkler system, which means the system is inactive. In some systems, water pressure from the municipal water supply may not be strong enough to push enough water to all parts of a building. In these cases, a fire pump is usually required. When the fire pump runs, a supervisory signal is sent indicating that the fire pump is activated. If the pump runs to maintain system pressure and does not shut down within a reasonable time, a site visit may be required. When water is scarce or unavailable, an on-site water tank may be required. Supervisory signals may be generated when the temperature of the water drops to a level low enough to freeze or if the water level or pressure drops below a safe level. In a dry system, a supervisory signal is generated if the air pressure drops below a usable level. If fire pump power is monitored, lack of power or power phase reversal will cause a trouble signal.

6.0.0 CONTROL PANELS

Today, there are many different companies manufacturing hundreds of different control panels, and more are developed each year. *Figure 28* shows a typical intelligent, addressable control panel. The control panel shown has a voice command center added for use in high-rise buildings. It provides for automatic evacuation messages, firefighter paging, and two-way communication to a central station through a telephone network. Although many control panels may have unique features, all control panels perform some specific basic functions. As shown in *Figure 29*, they detect problems through the sensor devices connected to

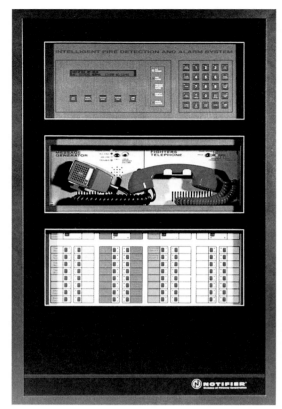

26405-14_F28.EPS

Figure 28   Typical intelligent, addressable control panel.

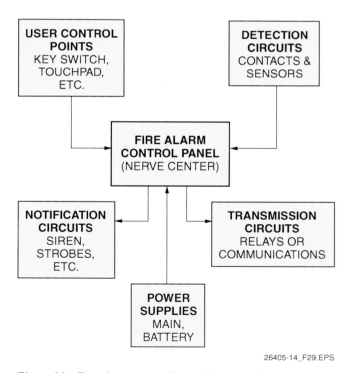

26405-14_F29.EPS

Figure 29   Fire alarm control panel inputs and outputs.

Water Flow Detectors

This water flow detector is typical of those used with wet-pipe sprinkler systems. Water flow through the associated pipe deflects the detector's vane. This activates the internal switch contacts, initiating an alarm or auxiliary indication. Water flow in the pipe can be caused by the opening of one or more sprinkler heads because of a fire, the opening of a test valve, or a leaking or ruptured pipe.

26405-14_SA06.EPS

them, and they sound alerts or report these problems to a central location.

A control panel can allow the user to reprogram the system and, in some cases, to activate or deactivate the system or zones. The user may also change the sensitivity of detectors in certain zones of the system. On panels that can be used for both intrusion and fire alarms, controls are provided to allow the user to enter or leave the monitored area without setting off the system. In addition, controls allow some portions of the system (fire, panic, holdup, etc.) to remain armed 24 hours a day, 365 days a year. Control panels also provide the alarm user, responding authority, or inspector with a way to silence bells or control other system features.

Control panels are usually equipped with LED, alphanumeric, or graphic displays to communicate information such as active zones or detectors to an alarm user or monitor. These types of displays are usually included on the FACP, but can also be located remotely if required.

The control panel organizes all the components into a working and functional system. The control panel is often referred to as the "brain" of the system. The control panel coordinates the actions the system takes in response to messages it gets from initiating devices or, in some cases, notification devices that are connected to it. Depending on the inputs from the devices, it may activate the notification devices and may also transmit data to a remote location via transmission circuits. In most cases, the control panel also conditions the power it receives from the building power system or from a backup battery so that it can be used by the fire alarm system.

6.1.0 User Control Points

User controls allow the alarm user to turn all or portions of the system on or off. User controls also allow the alarm user to monitor the system status, including the sensors or zones that are active, trouble reports, and other system parameters. They also allow the user to reset indicators of system events such as alarms or indications of trouble with phone lines, equipment, or circuits. In today's systems, many devices can be used to control the system, including keypads, key switches, touch screens, telephones (including wireless phones), and computers.

6.1.1 Keypads

Two general types of keypads are in use today: alphanumeric (*Figure 30*) and LED. Either type can be mounted at the control or at a separate location. Both allow the entry of a numerical code to program various system functions. An alphanumeric keypad combines a keypad that is similar to a pushbutton telephone dial with an alphanumeric display that is capable of showing letters and numbers. LED keypads use a similar keypad, but display information by lighting small LEDs.

6.1.2 Touch Screens

Touch screens allow the user to access multiple customized displays in graphic or menu formats. This enables rapid and easy interaction with the system.

6.1.3 Telephone/Computer Control

Since most fire alarm systems are connected to telephones for notification purposes, and telephones are often in locations where user control devices have traditionally been located, some control panel manufacturers have incorporated ways

26405-14_F30.EPS

Figure 30 Alphanumeric keypad and display.

for the telephone to function as a user interface. When telephones or cell phones are used as an interface for the system, feedback on system status and events is given audibly over the phone. Because the system is connected to the telephone network, a system with this feature can be monitored and controlled from anywhere a landline or cell phone call can be made. Most new systems can also be connected to a computer either locally or via a telephone modem or a high-speed data line to allow control of events and receipt of information.

6.2.0 FACP Initiating Circuits

An initiating circuit monitors the various types of initiating devices. A typical initiating circuit can monitor for three states: normal, trouble, and alarm. Initiating circuits can be used for monitoring fire devices, non-fire devices (supervisory), or devices for watch patrol or security. The signals from the initiating devices can be separately indicated at the premises and remotely monitored at a central station.

On Site

Intelligent Systems

An intelligent system uses analog devices to communicate with an intelligent, addressable control panel typical of the one shown in *Figure 28*. The control panel individually monitors the value or status reported by the analog sensors and makes normal, alarm, or trouble decisions. The specific control panel in *Figure 28* is a software-controlled unit capable of monitoring 1,980 individually identifiable and controllable detection/control points. It is field programmable from the front panel keyboard.

6.2.1 Initiating Circuit Zones

The term *zone* as it relates to modern fire alarm systems can have several definitions. Building codes restrict the location and size of a zone to enable emergency response personnel to quickly locate the source of an alarm. The traditional use of a zone in a conventional system was that each initiating circuit was a zone. Today, a single device or multiple initiating circuits or devices may occupy a zone. Generally, a zone may not cover more than one floor, exceed a certain number of square feet, or exceed a certain number of feet in length or width. In addition, some codes require different types of initiating devices to be on different initiating circuits within a physical area or zone. With modern addressable systems, each device is like a zone because each is displayed at the panel. It is common in addressable systems to group devices of several types into a zone. This is usually accomplished by programming the panel rather than by hardwiring the devices. Ultimately, the purpose of a zone is to provide the system monitor with information as to the location of a system alarm or problem. Local codes provide the specific guidelines.

6.2.2 Alarm Verification

To reduce false alarms, most FACPs allow for a delay in the activation of notification devices upon receiving an alarm signal from an initiating device. In some conventional systems using a positive alarm sequence feature, alarm delay is usually adjustable for a period up to three minutes, but must not exceed three minutes. This allows supervising or monitoring personnel to investigate the alarm. If a second detector activates during the investigation period, a fire alarm is immediately sounded. In other conventional systems, verification can be of the reset-and-resample type: The panel resets the detector and waits 10 to 20 seconds for the sensor to retransmit the alarm. If there is an actual fire, the sensor should detect it both times, and an alarm is then activated. In addressable systems, alarm verification can be of the wait-and-check type: the panel notes the initiation signal and then waits for about 10 to 20 seconds to see if the initiation signal remains constant before activating an alarm.

Another verification method for large areas with multiple sensors is to wait until two or more sensors are activated before an alarm is initiated. In this method, known as cross-zoning, a single detector activation may set a one minute or less pre-alarm condition warning signal at a manned monitoring location. If the pre-alarm signal is not cancelled within the allowed time, or if a second detector activates, a fire alarm is sounded. The pre-alarm signals are not required by the NFPA. In cross-zoning, two detectors must occupy each space regardless of the size of the space, and each detector may cover only half of the normal detection area. Cross-zoning along with other methods of alarm verification cannot be used on the same devices. Some building codes require alarm verification or some equivalent on all smoke detection devices.

Verification is commonly used in hotels, motels, hospitals, and other institutions with large numbers of detectors. It may also be used in monitored fire alarm systems for households to reduce false alarms.

6.2.3 FACP Labeling

Effective system design and installation requires that zones (or sensors) be labeled in a way that makes sense to all who will use or respond to the system. In some cases, this may require two sets of labels: one for the alarm user and another for the police and fire authorities. A central station operator should be aware of both sets of labels, since he or she will be talking to both the police and the alarm user. Labeling for the alarm user is done using names familiar to the user (Johnny's room, kitchen, master bedroom, and so on). Labeling for the police and fire authorities is done from the perspective of looking at the building from the outside (first floor east, basement rear, second floor west, and so on).

6.3.0 Types of FACP Alarm Outputs

Most panels provide one or more of the following types of outputs:

- *Relay or dry contacts* – Relay contacts or other dry contacts are electrically isolated from the circuit controlling them, which provides some protection from external spikes and surges. Additional protection is sometimes needed. Always check the contact ratings to determine how much voltage and current the contacts can handle. Check with the manufacturer to determine if this provides adequate protection.
- *Built-in siren drivers* – Built-in siren drivers use a transistor amplifier circuit to drive a siren speaker. Impedance must be maintained within the manufacturer's specifications. Voltage drop caused by long wire runs of small conductors can greatly reduce the output level of the siren.
- *Voltage outputs* – Voltage outputs are not isolated from the control, and some protection may be required. Spikes and surges such as those generated by solenoid bells or the back

EMF generated when a relay is de-energized can be a problem.

- *Open collector outputs* – Open collector outputs are outputs directly from a transistor and have limited current output. They are often used to drive a very low-current device or relay. Filtering may be required. Great caution should be used not to overload these outputs, as the circuit board normally has to be returned to the manufacturer if an overload of even a very short duration occurs.

6.4.0 FACP Listings

It is mandatory that fire alarm control panels be listed for the purpose for which they will be used. Common UL listings for FACPs are *UL 864* (fire alarm control panels) and *UL 985* (household fire warning system units).

In combination fire/burglary panels, fire circuits must be on or active at all times, even if the burglary control is disarmed or turned off. Some codes prohibit using combination intrusion and fire alarm control panels in commercial applications.

A fire alarm control panel listed for household use by the UL may never be installed in any commercial application unless it also has the appropriate commercial listing, has been granted equivalency by the authority having jurisdiction (AHJ) in writing, or is specifically permitted by a document that supersedes the reference standards. The NFPA defines a household as a single- or two-family residential unit, and this term applies to systems wholly within the confines of the unit. Except for monitoring, no initiation from, or notification to, locations outside the residence are permitted. Although an apartment building or condominium is considered commercial, a household system may be installed within an individual living unit. However, any devices that are outside the confines of the individual living unit, such as manual pull stations, must be connected to an FACP listed as commercial.

7.0.0 FACP PRIMARY AND SECONDARY POWER

Primary power for a fire alarm system and the FACP is normally a source of power provided by a utility company. However, primary power for systems and panels can also be supplied from emergency uninterruptible backup primary power systems. To avoid service interruption, a fire alarm system may be connected on the line side of the electrical main service disconnect switch. The circuit must be protected with a circuit breaker or fuse no larger than 20 amperes (A).

Secondary power for a fire alarm system can be provided by a battery or a battery with an approved backup generator. Secondary power must be immediately supplied to the fire alarm system in the event of a primary power failure. As defined by *NFPA 72®*, standby time for operation of the system on secondary power must be no less than 24 hours for central, local, proprietary, voice communications, and household systems and no less than 60 hours for auxiliary and remote systems. The battery standby time may be reduced if a properly configured battery-backup generator is available. During an alarm condition, the secondary power must operate the system under load for a minimum of 4, 5, or 15 minutes after expiration of the standby period, depending on the type of system. Any batteries used for secondary power must be able to be recharged within 48 hours.

8.0.0 NOTIFICATION APPLIANCES

Notification to building occupants of the existence of a fire is the most important life safety function of a fire alarm system. There are two primary types of notification: audible and visible. Concerns resulting from the Americans with Disabilities Act (ADA) have prompted the introduction of olfactory (sense of smell) and tactile (sense of touch) types of notification as well.

8.1.0 Visual Notification Devices

Strobes are high-intensity lights that flash when activated. They can be separate devices or mounted on or near the audible device. Strobe lights are more effective in residential, industrial, and office areas where they don't compete with other bright objects. Because ADA requirements dictate clear or white xenon (or equivalent) strobe lights, most interior visual notification devices are furnished with clear strobe lights (*Figure 31*). Strobe devices can be wall-mounted or ceiling-mounted and are usually combined with an audible notification device. When more than two strobes are visible from any location, the strobes must be synchronized to avoid random flashing, which can be disorienting and may actually cause seizures in certain individuals. Some strobes are brighter than others to accommodate different applications. These units may be rated in candelas.

8.2.0 Audible Notification Devices

Audible alarm devices are noise-making devices, such as sirens, bells, or horns, that are used as part of a local alarm system to indicate an alarm condition. In some cases, low-current chimes or

CEILING-MOUNTED   WALL-MOUNTED

26405-14_F31.EPS

Figure 31   Typical ceiling-mounted and wall-mounted strobe devices.

On Site

Audible Alarms

Audible alarms must have an intensity and frequency that attracts the attention of those with partial hearing loss. Such alarms shall produce a sound that exceeds the prevailing sound level in the space by 15 audible decibels (dBA) or exceeds the maximum sound level with a duration of 60 seconds by 5dBA, whichever is louder. Sound levels shall not exceed 120dBA.

Some jurisdictions prohibit the use of bell notification devices in schools as they can be confused with classroom bells.

buzzers are also used. Audible signals used for fire alarms in a facility that contains other sound-producing devices must produce a unique sound pattern so that a fire alarm can be recognized. If more than one audible signal is used in a facility, they must be synchronized to maintain any sound pattern. *NFPA 72®* specifies that a standard signal known as Temporal three is required by most fire alarm systems, including household systems. This signal has three ½-second tones, a pause, and then a repeat of the pattern until the alarm is manually reset. Only a fire alarm may use this signal.

Some bells (*Figure 32*) use an electrically-vibrated clapper to repeatedly strike a gong. These types of solenoid-operated bells can also produce electrical noise, which will interfere with controls unless proper filters are used. Solenoid-operated bells draw relatively high current, typically 750 to 1,500 milliamps (mA). Another type of bell is a motor-driven bell. In this type of bell, a motor drives the clapper and produces louder sounds than a solenoid-operated bell. Electrical interference is eliminated and less current is required to operate the bell.

Self-contained sirens are combinations of speakers and sound equipment. They produce siren sounds and can be used for voice

announcements. Self-contained siren packaging saves installation time. If more than one siren is required, they must be synchronized.

A horn (*Figure 33*) usually consists of a continuously vibrating membrane or a piezoelectric element. In a supervised notification circuit, all devices are polarized, allowing current to flow in one direction only. A buzzer uses less power and consists of a continuously vibrating membrane like a horn. Chimes are electronic devices, like self-contained sirens, and have a very low current draw.

On Site

Strobe Lights

Strobe lights may draw a relatively high current compared to other types of notification devices. For this reason, they may require a separate circuit.

26405-14_F32.EPS

Figure 32   Typical bell with a strobe light.

Figure 33 Typical horn with a strobe light.

26405-14_F34.EPS

Figure 34 Voice evacuation system.

8.3.0 Voice Evacuation Systems

With a voice evacuation system (*Figure 34*), building occupants can be given instructions in the event of an emergency. A voice evacuation system can be used in conjunction with an FACP or as a stand-alone unit with a built-in power supply and battery charger. Voice announcements (*Figure 35*) can be made to inform occupants what the problem is or how to evacuate. In many cases, the voice announcements are prerecorded and selected as required by the system. The announcements can also be made from a microphone (*Figure 36*) located at the FACP or at a remote panel. *Figure 36* shows several types of speakers used for voice evacuation systems.

Temporal three signaling is not used with zoned voice evacuation systems.

8.4.0 Signal Considerations

Closed doors may drop the audible decibel (dBA) sound levels of alarms below those required to wake children or hearing-impaired adults. Air conditioners, room humidifiers, and other equipment may cause noise levels to increase. Also, sirens, horns, and bells may fail or be disabled by the fire. Multiple fire sounders provide redundancy and, if properly placed, will provide ample decibel levels to wake all sleeping occupants.

The following is a description of the types of code-authorized signals supplied to various notification devices from a typical fire control panel:

- *General signal* – General signals operate throughout the entire building. Evacuation signals require a distinctive signal. The code requires that a Temporal three signal pattern be used that is in accordance with *ANSI Standard S3.41* (and *ISO 8201*), *NFPA 101*®, and *NFPA 72*®. Temporal three signals consist of three short ½-second tones with ½-second pauses between the tones. This is followed by a 1-second silent period, and then the process is repeated.

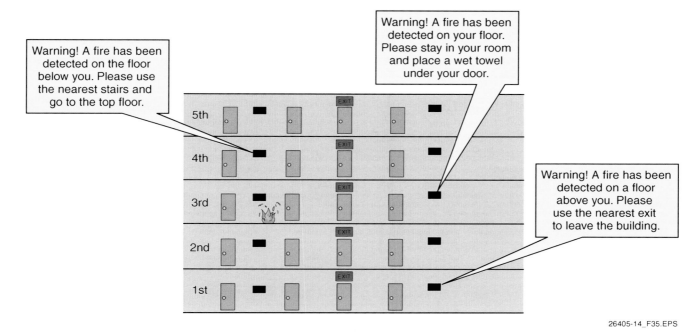

Warning! A fire has been detected on the floor below you. Please use the nearest stairs and go to the top floor.

Warning! A fire has been detected on your floor. Please stay in your room and place a wet towel under your door.

Warning! A fire has been detected on a floor above you. Please use the nearest exit to leave the building.

26405-14_F35.EPS

Figure 35   Typical voice evacuation messages.

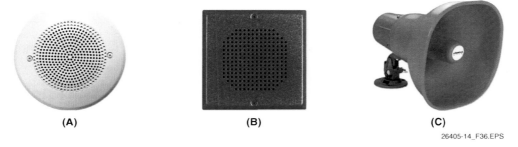

(A)   (B)   (C)

26405-14_F36.EPS

Figure 36   Typical speakers.

• *Attendant signal* – Attendant signaling is used where assisted evacuation is required due to such factors as age, disability, and restraint. Such systems are commonly used in conjunction with coded chimes throughout the area or a coded voice message to advise staff personnel of the location of the alarm source. A coded message might be announced such as, "Mr. Green, please report to the third floor nurses' station." No general alarm needs to be sounded, and no indicating appliances need to be activated throughout the area. The attendant signal feature requires the approval of the AHJ.

• *Pre-signal* – A pre-signal is an attendant signal with the addition of human action to activate a general signal. It is also used when the control delays the general alarm by more than one minute. The pre-signal feature requires the approval of the AHJ. A signal to remote locations must activate upon an initial alarm signal.

• *Positive alarm sequence* – When an alarm is initiated, the general alarm is not activated for 15 seconds. If a staff person acknowledges the alarm within the 15-second window, the general alarm is delayed for three minutes so that the staff can investigate. Failure to manually acknowledge the initial attendant signal will automatically cause a general alarm. Failure to abort the acknowledged signal within 180 seconds will also automatically cause a general alarm. If any second selected automatic detector is activated during the delay window, the system will immediately cause a general alarm. An activation of a manual station will automatically cause a general alarm. This is an excellent technique for the prevention of unwanted general alarms; however, trained personnel are an integral part of these systems. The positive alarm sequence feature requires the approval of the AHJ.

- *Voice evacuation* – Voice evacuation may be either live or prerecorded, and automatically or manually initiated. Voice evacuation systems are a permitted form of general alarm and are required in certain occupancies, as specified in the applicable chapter of *NFPA 101*. Voice evacuation systems are often zoned so that only the floors threatened by the fire or smoke are immediately evacuated. Zoned evacuation is used where total evacuation is physically impractical. Temporal three signaling is not used with zoned evacuation. Voice evacuation systems must maintain their ability to communicate even when one or more zones are disabled due to fire damage. Buildings with voice evacuation usually include a two-way communication system (fireman's phone) for emergency communications. High-rise buildings are required to have a voice evacuation system.
- *Alarm sound levels* – Where a general alarm is required throughout the premises, audible signals must be clearly heard above maximum ambient noise under normal occupancy conditions. Public area audible alarms should be 75dBA at 10' and a maximum of 120dBA at minimum hearing distance. Public area audible alarms must be at least 15dBA above ambient sound level, or a maximum sound level that lasts over 60 seconds measured at 5' above finished floor (AFF) level. Audible alarms to alert persons responsible for implementing emergency plans (guards, monitors, supervisory personnel, or others) must be between 45dBA and 120dBA at minimum hearing distance. If an average sound level greater than 105dBA exists, the use of a visible signal is required. Typical average ambient sound levels from *NFPA 72* are given in *Table 2*.

NOTE

The data in *Table 2* is merely a guide. Actual conditions may vary.

Closed doors may drop the dB levels below those required to wake children or the hearing-impaired. Air conditioners, room humidifiers, and similar equipment may cause the noise levels to increase (55dBA typical). Sirens, horns, bells, and other indicating devices (appliances) may fail or be disabled by a fire. To reduce the effect of these disabled devices, multiple fire sounders should be considered. This may also help in providing ample dB levels to wake sleeping occupants throughout the structure (70dBA at the pillow is commonly accepted as sufficient to wake a sleeping person).

All residential alarm sounding devices must have a minimum dB rating of 85dBA at 10'. An exception to this rule is when more than one sounding device exists in the same room. A sounding device in the room must still be 85dBA at 10', but all additional sounding devices in that room may have a rating as low as 75dBA at 10'. Sound intensity doubles with each 3dB gain and is reduced by one-half with each 3dB loss. Doubling the distance to the sound source will cause a 6dB loss. Some other loss considerations are given in *Table 3*.

- *Coded versus non-coded signals* – A coded signal is a signal that is pulsed in a prescribed code for each round of transmission. For example, four pulses would indicate an alarm on the fourth floor. Temporal three is not a coded signal and is only intended to be a distinct, general fire alarm signal. A non-coded signal is a signal

Table 2 Typical Average Ambient Sound Levels

Area	Sound Level (dBA)	Area	Sound Level (dBA)
Mechanical rooms	85	Educational occupancies	45
Industrial occupancies	80	Underground structures	40
Busy urban thoroughfares	70	Windowless structures	40
Urban thoroughfares	55	Mercantile occupancies	40
Institutional occupancies	50	Places of assembly	40
Vehicles and vessels	50	Residential occupancies	35
Business occupancies	45	Storage	30

Table 3 Typical Sound Loss at 1,000Hz

Area	Loss (dBA)
Stud wall	41
Open doorway	4
Typical interior door	11
Typical fire-rated door	20
Typical gasketed door	24

that is energized continuously by the control. It may pulse, but the pulsing will not be designed to indicate any code or message.

- *Visual appliance signals* – Notification signals for occupants to evacuate must be by audible and visible signals in accordance with *NFPA 72®* and *CABO/ANSI A117.1.* However, there may or may not be exceptions to this rule. Under the existing *NFPA 101®* building chapter, only audible signals are required in premises where:
  - No hearing-impaired occupant is ever present under normal operation
  - In hotels and apartments where special rooms are made available to the hearing-impaired
  - Where the AHJ approves alternatives to visual signals (ADA codes may or may not allow these exceptions)

9.0.0 COMMUNICATIONS AND MONITORING

Communications is a means of sending information to personnel who are too far away to directly see or hear a fire alarm system's notification devices. It is the transmission and reception of information from one location, point, person, or piece of equipment to another. Understanding the information that a fire alarm system communicates makes it easier to determine what is happening at the alarm site. Knowing how that information gets from the alarm site to the monitoring site is helpful if a problem occurs somewhere in between.

9.1.0 Monitoring Options

There are several options for monitoring the signals of an alarm system. They include:

- *Central station* – A location, normally run by private individuals or companies, where operators monitor receiving equipment for incoming fire alarm system signals. The central station

may be a part of the same company that sold and installed the fire alarm system. It is also common for the installing company to contract with another company to do the monitoring on its behalf.

- *Proprietary* – A facility similar to a central station except that the notification devices are located in a constantly staffed room maintained by the property owner for internal safety operations. The personnel may respond to alarms, alert local fire departments when alarms are activated, or both.
- *Certified central station* – Monitoring facilities that are constructed and operated according to a standard and are inspected by a listing agency to verify compliance. Several organizations, including the UL, publish criteria and list those central stations that conform to those criteria.

9.2.0 Digital Communicators

Digital communicators use standard telephone lines or wireless telephone service to send and receive data. Costs are low using this method because existing voice lines may be used, eliminating the need to purchase additional communication lines. Standard voice-grade telephone lines are also easier to repair than special fire alarm communication lines.

Digital communicators are connected to a standard, voice-grade telephone line through a special connecting device called the RJ31-X (*Figure 37*). The RJ31-X is a modular telephone jack into which a cord from the digital communicator is plugged. The RJ31-X separates the telephone company's equipment from the fire alarm system equipment and is approved by the Federal Communications Commission (FCC).

Although using standard telephone lines has several advantages, problems may arise if the customer and the fire alarm system both need the phone at the same time. In the event of an alarm, a technique known as line seizure gives the fire alarm system priority. The digital communicator is connected to the phones and can control or seize the line whenever it needs to send a signal. If the customer is using the telephone when the alarm system needs to send a signal, the digital communicator will disconnect the customer until the alarm signal has been sent. Once the signal is sent, the customer's phones are reconnected.

The typical sequence that occurs when the digital communicator for an alarm system is activated is shown in *Figure 38*. When an alarm is to be sent,

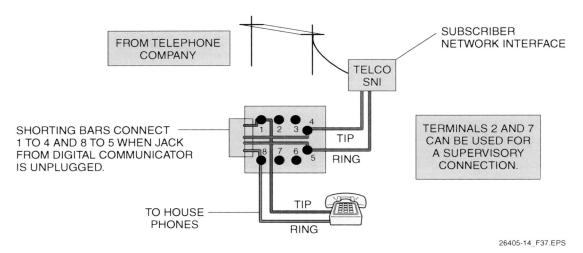

Figure 37 RJ31-X connection device.

FROM TELEPHONE
COMPANY

SUBSCRIBER
NETWORK INTERFACE

TELCO
SNI

SHORTING BARS CONNECT
1 TO 4 AND 8 TO 5 WHEN JACK
FROM DIGITAL COMMUNICATOR
IS UNPLUGGED.

TIP

RING

TERMINALS 2 AND 7
CAN BE USED FOR
A SUPERVISORY
CONNECTION.

TO HOUSE
PHONES

TIP

RING

26405-14_F37.EPS

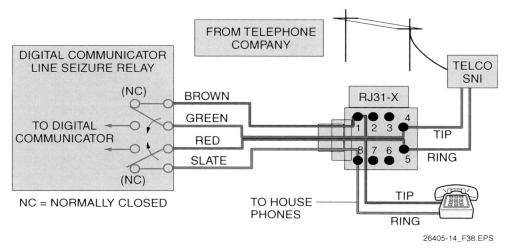

Figure 38 Line seizure.

DIGITAL COMMUNICATOR
LINE SEIZURE RELAY

(NC)

BROWN

FROM TELEPHONE
COMPANY

TELCO
SNI

GREEN

TO DIGITAL
COMMUNICATOR

RED

RJ31-X

TIP

SLATE

RING

(NC)

NC = NORMALLY CLOSED

TO HOUSE
PHONES

TIP

RING

26405-14_F38.EPS

the digital communicator energizes the seizure relay. The activated relay disconnects the house phones and connects the communicator to the Telco line. After the communicator detects a signal called the kiss-off tone, it de-energizes the seizure relay. This disconnects the communicator and reconnects the house phones.

In combination fire and security systems, the fire alarm supersedes the security alert. The RJ31-X will send the fire alarm signal before sending the security alert signal.

The RJ31-X is a modular connection and, like a standard telephone cord, it can unplug easily. If the cord remains unplugged from the jack, the digital communicator will be disconnected from the telephone line until the cord is reconnected. This creates a problem because the digital communicator cannot reach the digital receiver without the telephone line. This reduces the effectiveness of the alarm system, even though local notification devices are activated.

The NFPA uses the following terms to refer to digital communications:

- Digital Alarm Communicator Receiver (DACR) – This is a system component that will accept and display signals from Digital Alarm Communicator Transmitters (DACT) sent over the Public Switched Telephone Network.
- Digital Alarm Communicator System (DACS) – This is a system in which signals are transmitted from a DACT (located in the secured area) through the public switched telephone network to a DACR.
- *Digital Alarm Communicator Transmitter (DACT)* – This is a device that sends signals over the public switched telephone network to a DACR.

The following conditions have been established by *NFPA 72*® with regard to digital communicators:

- They can be used as a remote supervising station fire alarm system when acceptable to the AHJ.
- Only loop start and not ground start lines can be used.
- The communicator must have line seizure capability.
- A failure-to-communicate signal must be shown if ten attempts are made without getting through.
- They must connect to two separate phone lines at the protected premises. Exception: The secondary line may be a radio system (this does not apply to household systems).
- Failure of either phone line must be annunciated at the premises within four minutes of the failure.
- If long distance telephone service, including Wide Area Telephone Service (WATS) is used, the second telephone number shall be provided by a different long distance provider, where available.
- Each communicator shall initiate a test call to the central station at least once every 24 hours (this does not apply to household systems).

Telephone Line Problems

The RJ31-X can be helpful in determining the nature of a problem with the telephone line. One method of determining if the fire alarm system has caused a problem with the telephone lines is to unplug the cord to the RJ31-X. This disconnects the fire alarm system and restores the connections of the telephone line. If the customer's telephones function properly when the cord is unplugged, the fire alarm system may be causing the problem. If the problem with the telephone line persists with the cord to the RJ31-X unplugged, the source of the problem is not the alarm system.

9.3.0 Cellular Backup

Some fire alarm systems that rely on phone lines for communications have a cellular backup system that utilizes wireless phone technology to restore communications in the event of a disruption in normal telephone line service (*Figure 39*).

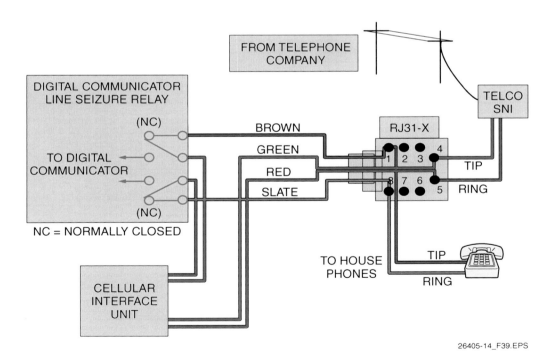

26405-14_F39.EPS

Figure 39   Cellular backup system.

10.0.0 GENERAL INSTALLATION GUIDELINES

This section contains general installation information applicable to all types of fire alarm systems. For specific information, always refer to the manufacturer's instructions, the building drawings, and all applicable local and national codes.

10.1.0 General Wiring Requirements

NEC Article 760 specifies the wiring methods and special cables required for fire protective signaling systems. The following special cable types are used in protective signaling systems:

* Power-limited fire alarm (FPL) cable
* Power-limited fire alarm riser (FPLR) cable
* Power-limited fire alarm plenum (FPLP) cable
* Nonpower-limited fire alarm (NPLF) circuit cable
* Nonpower-limited fire alarm riser (NPLFR) circuit riser cable
* Nonpower-limited fire alarm plenum (NPLFP) circuit cable

In addition to *NEC Article 760*, the following *NEC®* articles cover other items of concern for fire alarm system installations.

* *NEC Sections 110.11 and 300.6(A), (B), and (C), Corrosive, Damp, or Wet Locations*
* *NEC Section 300.21, Spread of Fire or Products of Combustion*
* *NEC Section 300.22, Ducts, Plenums, and Other Air Handling Spaces*
* *NEC Articles 500 through 516 and 517, Part IV, Locations Classified as Hazardous*
* *NEC Article 695, Fire Pumps*
* *NEC Article 725, Remote-Control and Signaling Circuits (Building Control Circuits)*
* *NEC Article 770, Fiber Optics*
* *NEC Article 800, Communications Circuits*
* *NEC Article 810, Radio and Television Equipment*

In addition to the *NEC®*, some AHJs may specify requirements that modify or add to the *NEC®*. It is essential that a person or firm engaged in fire alarm work be thoroughly familiar with the *NEC®* requirements, as well as any local requirements for fire alarm systems.

10.2.0 Workmanship

Fire alarm circuits must be installed in a neat and workmanlike manner. Cables must be supported by the building structure in such a manner that the cables will not be damaged by normal building use. One way to determine accepted industry practice is to refer to nationally recognized standards such as *Commercial Building Telecommunications Wiring Standard, ANSI/EIA/TIA 568; Commercial Building Standard for Telecommunications Pathways and Spaces, ANSI/EIA/TIA 569;* and *Residential and Light Commercial Telecommunications Wiring Standard, ANSI/EIA/TIA 570.*

10.3.0 Access to Equipment

Access to equipment must not be blocked by an accumulation of wires and cables that prevents removal of panels, including suspended ceiling panels.

10.4.0 Fire Alarm Circuit Identification

Fire alarm circuits must be identified at the control and at all junctions as fire alarm circuits. Junction boxes must be clearly marked as fire junction boxes to prevent confusion with commercial light and power. AHJs differ on what constitutes clear marking. Check the requirements before starting work. The following are examples of some AHJ-acceptable markings:

* Red painted cover
* The words Fire Alarm on the cover
* Red painted box
* The word Fire on the cover
* Red stripe on the cover

10.5.0 Power-Limited Circuits in Raceways

Power-limited fire circuits must not be run in the same cable, raceway, or conduit as high-voltage circuits. Examples of high-voltage circuits are electric light, power, and nonpower-limited fire (NPLF) circuits. When NPLF cables must be run in the same junction box, they must be run in accordance with the *NEC®*, including maintaining a ¼" spacing from Class 1, power, and lighting circuits.

10.6.0 Mounting of Detectors

Observe the following precautions when mounting detectors:

* Circuit conductors are not supports and must not be used to support the detector.
* Plastic masonry anchors should not be used for mounting detectors to gypsum drywall, plaster, or drop ceilings.
* Toggle or winged expansion anchors should be the minimum used for gypsum drywall.
* The best choice for mounting is an electrical box. Most equipment is designed to be fastened to a standard electrical box.

All fire alarm devices should be mounted to the appropriate electrical box as specified by the manufacturer. If not mounted to an electrical box, fire alarm devices must be mounted by other means as specified by the manufacturer. The use of plastic masonry anchors does not mean that a plastic anchor designed for use in drywall cannot be used in drywall. Masonry plastic anchors do not open as far on the end as drywall plastic anchors. Always read and follow the manufacturer's instructions; doing so is necessary to meet the requirements of the UL listing.

10.7.0 Outdoor Wiring

Fire alarm circuits extending beyond one building are governed by the *NEC*®. The *NEC*® sets the standards on the size of cable and methods of fastening required for cabling. Some manufacturers prohibit any aerial wiring. The *NEC*® also specifies clearance requirements for cable from the ground. Overhead spans of open conductors and open multi-conductor cables of not over 600V, nominal, must be at least 10' (3.05 m) above finished grade, sidewalks, or from any platform or projection from which they might be reached, where the supply conductors are limited to 150V to ground and accessible to pedestrians only. Additional requirements apply for areas with vehicle traffic. In addition, the *NEC*® states that fire alarm wiring can be attached to the building, but must be protected against physical damage as afforded by baseboards, door frames, ledges, etc. per *NEC Section 760.130(B)(1)*.

10.8.0 Fire Seals

Electrical equipment and cables must not be installed in a way that might help the spread of fire. The integrity of all fire-rated walls, floors, partitions, and ceilings must be maintained. An approved sealant or sealing device must be used to fill all penetrations. Any wall that extends from the floor to the roof or from floor-to-floor should be considered a firewall. In addition, raceways and cables that go from one room to another through a fire barrier must be sealed.

10.9.0 Wiring in Air Handling Spaces

Wiring in air handling spaces requires the use of approved wiring methods, including:

- Special plenum-rated cable
- Flexible metal tubing (Greenfield)
- Electrical metallic tubing (EMT)
- Intermediate metallic conduit (IMC)
- Rigid metallic conduit (hard wall or Schedule 80 conduit)

Standard cable tie straps are not permissible in plenums and other air handling spaces. Ties must be plenum rated. Bare solid copper wire used in short sections as tie wraps may be permitted by most AHJs. Fire alarm equipment is permitted to be installed in ducts and plenums only to sense the air. All splices and equipment must be contained in approved fire-resistant and low-smoke-producing boxes.

10.10.0 Wiring in Hazardous Locations

The *NEC®* includes requirements for wiring in hazardous locations. Some areas that are considered hazardous are listed below:

- *NEC Article 511, Commercial Garages, Repair and Storage*
- *NEC Article 513, Aircraft Hangars*
- *NEC Article 514, Motor Fuel Dispensing Facilities*
- *NEC Article 515, Bulk Storage Plants*
- *NEC Article 516, Spray Application, Dipping, and Coating*
- *NEC Article 517, Healthcare Facilities*
- *NEC Article 518, Assembly Occupancies*
- *NEC Article 520, Theaters and Similar Locations*
- *NEC Article 545, Manufactured Buildings*
- *NEC Article 547, Agricultural Buildings*

10.11.0 Remote Control Signaling Circuits

Building control circuits (*Figure 40*) are normally governed by *NEC Article 725*. However, circuit wiring that is both powered and controlled by the fire alarm system is governed by *NEC Article 760*. A common residential problem occurs when using a system-type fire alarm that is a combination burglar and fire alarm. Many believe that the keypad is only a burglar alarm device. If the keypad is also used to control the fire alarm, it is a fire alarm device, and the cable used to connect it to the control panel must comply with *NEC Article 760*. Also, motion detectors that are controlled and powered by the combination fire alarm and burglar alarm power must be wired in accordance with *NEC Article 760* (fire-rated cable).

10.12.0 Cables Running Floor to Floor

Riser cable is required when wiring runs from floor to floor. The cable must be labeled as passing a test to prevent fire from spreading from floor to floor. An example of riser cable is FPLR. This requirement does not apply to one- and two-family residential dwellings.

10.13.0 Cables Running in Raceways

All cables in a raceway must have insulation rated for the highest voltage used in the raceway. Power-limited wiring may be installed in raceways or conduit, exposed on the surface of a ceiling or wall, or fished in concealed spaces. Cable splices or terminations must be made in listed fittings, boxes, enclosures, fire alarm devices, or utilization equipment. All wiring must enter boxes through approved fittings and be protected against physical damage.

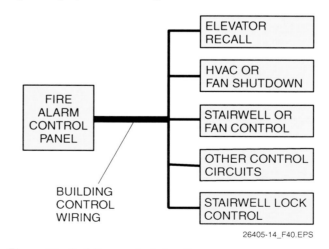

Figure 40 Building control circuits.

10.14.0 Cable Spacing

Power-limited fire alarm circuit conductors must be separated at least 2" (50.8 mm) from any electric light, power, Class 1, or nonpower-limited fire alarm circuit conductors. This is to prevent damage to the power-limited fire alarm circuits from induced currents caused by the electric light, power, Class 1, or nonpower-limited fire alarm circuits.

10.15.0 Elevator Shafts

Wiring in elevator shafts must directly relate to the elevator and be installed in rigid metallic conduit, rigid nonmetallic conduit, EMT, IMC, or up to 6' of flexible conduit.

10.16.0 Terminal Wiring Methods

The wiring for circuits using EOL terminations must be done so that removing the device causes a trouble signal (*Figure 41*).

10.17.0 Conventional Initiation Device Circuits

There are three styles of Class B and two styles of Class A conventional initiation device circuits listed in *NFPA 72®*. Various local codes address what circuit types are required. *NFPA 72®*

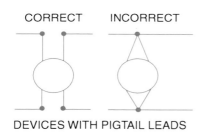

DEVICES WITH PIGTAIL LEADS

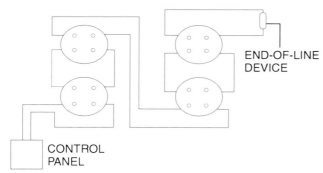

26405-14_F41.EPS

Figure 41   Correct wiring for devices with EOL terminations.

describes how they are to operate. A brief explanation of some of the circuits follows:

- *Class B, Style A* – Fire alarm control panels (FACPs) using this style are no longer made in the United States, but a few systems remain in operation. Single (wire) opens and ground faults are the only types of trouble that can be indicated, and the system will not receive an alarm in a ground fault condition. An alarm is initiated with a wire-to-wire short.

- *Class B, Style B* – (See *Figure 42*.) The FACP is required to receive an alarm from any device up to a break with a single open. An alarm is initiated with a wire-to-wire short. A trouble signal is generated for a circuit ground or open using an end-of-line (EOL) device that is usually a resistor. An alarm can also be received with a single ground fault on the system.

- *Class B, Style C* – (See *Figure 43*.) This style, while used in the United States, is more common in Europe. An open circuit, ground, or wire-to-wire short will cause a trouble indication. Devices or detectors in this type of circuit require a device (normally a resistor) in series with the contacts in order for the panel to detect an alarm condition. The panel will receive an alarm signal with a single ground fault on the system. The current-limiting resistor is normally lower in resistance than the end-of-line resistor.

- *Class A, Style D* – (See *Figure 44*.) In this type of circuit, an open or ground will cause a trouble signal. Shorting across the initiation loop will cause an alarm. Activation of any initiation device will result in an alarm, even when a single break or open exists anywhere in the circuit, because of the back loop circuit. The loop is returned to a special condition circuit, so there is no end of line and, therefore, no EOL device.

- *Class A, Style E* – (See *Figure 45*.) This style is an enhanced version of Style D. An open circuit, ground, or wire-to-wire short is a trouble condition. All devices require another device (normally a resistor) in series with the contacts to generate an alarm. Activation of any of the initiating devices will result in an alarm even if the initiation circuit has a single break or the system has a single ground.

The style of circuit can affect the following:

- The maximum quantity of each type of device permitted on each circuit
- The maximum quantity of circuits allowed for a fire alarm control panel/communicator

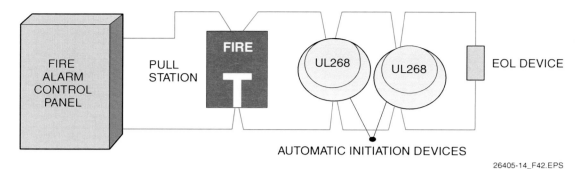

Figure 42   Typical Class B, Style B initiation circuit.

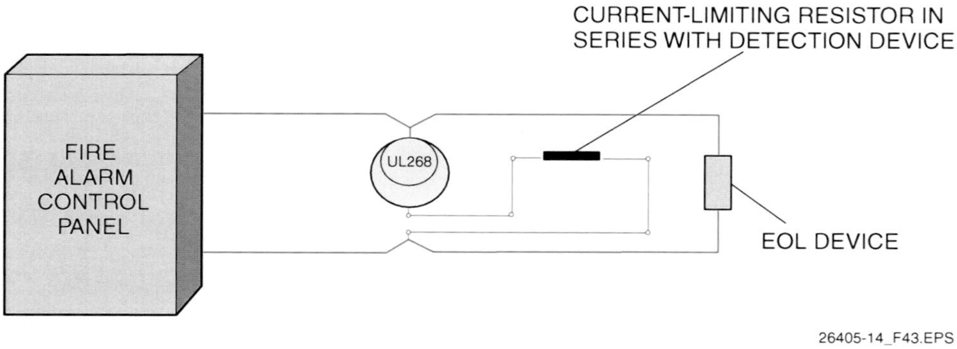

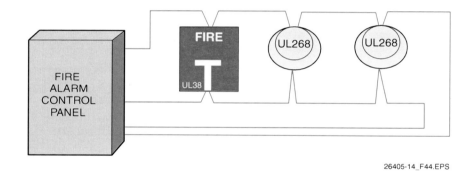

Figure 43   Typical Class B, Style C initiation circuit.

Figure 44   Typical Class A, Style D initiation circuit.

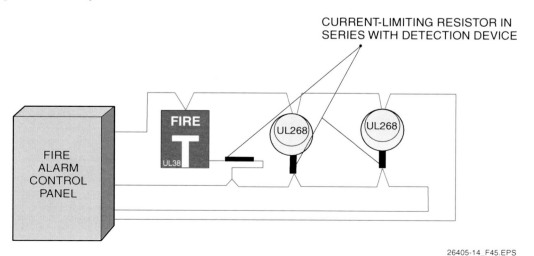

Figure 45   Typical Class A, Style E initiation circuit.

- The maximum quantity of buildings allowed for a signaling line circuit (SLC)
- The maximum quantity of signaling circuits and buildings allowed for a monitoring station

10.18.0 Notification Appliance Circuits

The following classes and styles of circuits are used for notification circuits (*Figure 46*):

- *Class B, Style W* – Class B, Style W is a two-wire circuit with an end-of-line device. Devices will operate up to the location of a fault. A ground may disable the circuit.
- *Class B, Style X* – Class B, Style X is a four-wire circuit. It has alarm capability with a single open, but not during a ground fault.
- *Class B, Style Y* – Class B, Style Y is a two-wire circuit with an end-of-line device. Devices on this style of circuit will operate up to the location of a fault. Ground faults are indicated differently than other circuit troubles.
- *Class A, Style Z* – Class A, Style Z is a four-wire circuit. All devices should operate with a single ground or open on the circuit. Ground faults are indicated differently than other circuit troubles.

Style X is similar to Style Z, except that during a ground fault, Style X will not operate and the panel will not be able to tell what type of trouble exists. Style W is similar to Style Y, except that during a ground fault, Style W will not operate and the panel will not be able to tell what type of

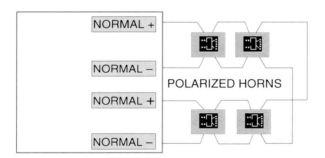

CLASS B, STYLE X OR CLASS A, STYLE Z

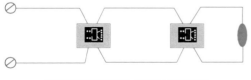

NOTIFICATION APPLIANCES

CLASS B, STYLES W AND Y

26405-14_F46.EPS

Figure 46 Typical notification appliance circuits.

Network Command Center

This PC-based network command center is used to display event information from local or wide area network devices in a text or graphic format. When a device initiates an alarm, the appropriate graphic floor plan is displayed along with operator instructions. Other capabilities of this command center include event history tracking and fire panel programming/control.

26405-14_SA08.EPS

trouble exists. Only Style Z operates all devices with a single open or a single ground fault. Only Style Z is a Class A notification appliance circuit (NAC).

All styles of circuits will indicate a trouble alarm at the premises with a single open and/or a single ground fault. Styles Y and Z have alarm capability with a single ground fault. Styles X and Z have alarm capability with a single open. In Class B circuits, the panel monitors whether or not the wire is intact by using an EOL device (wire supervision). Electrically, the EOL device must be at the end of the indicating circuit; however, Style X is an exception. Examples of EOL devices are resistors to limit current, diodes for polarity, and capacitors for filtering.

10.19.0 Primary Power Requirements

If more power is needed than can be supplied by one 20A circuit, additional circuits may be used, but none may exceed 20A. To help prevent system damage or false alarms caused by electrical surges or spikes, surge protection devices should be installed in the primary power circuits unless the fire alarm equipment has self-contained surge protection. In addition, some jurisdictions may require breaker locks and/or other means of identifying circuit breakers for FACPs.

10.20.0 Secondary Power Requirements

An approved generator supply or backup batteries must be used to supply the secondary power of a fire alarm system. The secondary power system must, upon loss of primary power, immediately keep the fire alarm functioning for at least as long as indicated in *Table 4*.

11.0.0 TOTAL PREMISES FIRE ALARM SYSTEM INSTALLATION GUIDELINES

This section covers the requirements for the proper installation, testing, and certification of fire alarm systems and related systems for totally protected premises.

11.1.0 Manual Fire Alarm Box (Pull Station) Installation

The following guidelines apply to the installation of a manual fire alarm box (pull station):

- Manual pull stations must be *UL 38*-listed or the equivalent. Multi-purpose keypads cannot be used as fire alarm manual pull stations unless *UL 38*-listed for that purpose.

- Manual pull stations must be installed in the natural path of escape, near each required exit from an area, in occupancies that require manual initiation. Ideally, they should be located near the doorknob edge of the exit door. In any case, they must be no more than 5' from the exit (*Figure 47*). In most cases, manual pull stations are installed with the actuators at the heights of 42" to 54" to conform to local ADA requirements. Most new pull stations are supplied with Grade II Braille on them for the visually impaired.

- The force required to operate manual pull stations must be no more than 5 foot-pounds.

- A manual pull station must be within 200' of horizontal travel on the same floor from any part of the building (*Figure 48*). If the distance is exceeded, additional pull stations must be installed on the floor.

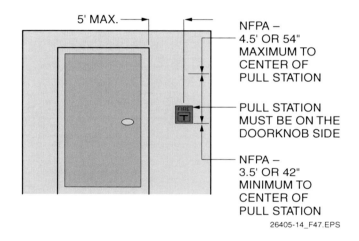

Figure 47 Pull station location and mounting height.

Table 4 Secondary Power Duration Requirements

NFPA Standard	Maximum Normal Load	Maximum Alarm Load
Central station	24 hours	See local system
Local system	24 hours	5 minutes
Auxiliary systems	60 hours	5 minutes
Remote stations	60 hours	5 minutes
Proprietary systems	24 hours	5 minutes
Household system	24 hours	4 minutes
Emergency voice alarm communications systems	24 hours	15 minutes maximum load 2 hours emergency operation

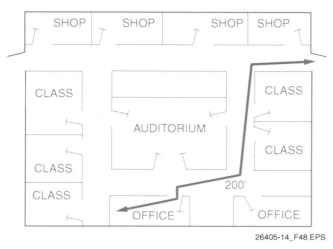

26405-14_F48.EPS

Figure 48 Maximum horizontal pull station distance from an exit.

11.2.0 Flame Detector Installation

When installing UV or IR flame detectors, the manufacturer's instructions and *NFPA 72®* should be consulted. The following should be observed when installing these detectors:

- *UV flame detectors*
  - Response is based on the distance from the fire, angle of view, and fire size.
  - While some units have a 180° field of view, sensitivity drops substantially with angles of more than 45° to 50°. Normally, the field of view is limited to less than 90° (*Figure 49*).
  - If used outdoors, make sure the unit is listed for outdoor use.
  - A UV detector is considered solar blind, but in order to prevent false alarms, it should never be aimed near or directly at any path that the sun can take.
  - Never aim the units into areas where electric arc welding or cutting may be performed.
- *IR flame detectors*
  - Response varies depending on the angle of view. At 45°, the sensitivity drops to 60% of the 0° sensitivity (*Figure 50*).
  - IR detectors cannot be used to detect alcohol, liquefied natural gas, hydrogen, or magnesium fires.
  - IR detectors work best in low light level installations. High light levels desensitize the units. Discriminating units can tolerate up to ten footcandles of ambient light. Non-discriminating units can tolerate up to two footcandles of ambient light.
  - Never use the units outdoors.

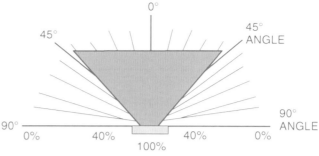

PERCENT OF RELATIVE SENSITIVITY

26405-14_F49.EPS

Figure 49 UV detector response.

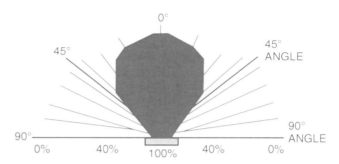

PERCENT OF RELATIVE SENSITIVITY

26405-14_F50.EPS

Figure 50 IR detector response.

11.3.0 Smoke Chamber Definition, Smoke Spread Phenomena, and Stratification Phenomena

The following sections cover smoke chamber definition, smoke spread phenomena, and stratification phenomena.

11.3.1 Smoke Chamber Definition

Before automatic smoke detectors can be installed, the area of coverage known as the smoke chamber must be defined. The smoke chamber is the continuous, smoke-resistant, perimeter boundary of a room, space, or area to be protected by one or more automatic smoke detectors between the upper surface of the floor and the lower surface of the ceiling. For the purposes of determining the area to be protected by smoke detectors, the smoke chamber is not the same as a smoke-tight compartment. It should be noted that some rooms that may have a raised floor and a false ceiling actually have three smoke chambers: the chamber beneath a raised floor, the chamber between the raised floor and the visible ceiling above, and the chamber between the room's visible ceiling and the floor above (or the lower portion of the roof). Few cases require detection in all these areas. However, some computer equipment rooms with a great deal of electrical power and communications cable under the raised floor may require detection in the chamber under the floor in addition to the chamber above the raised floor.

The simplest example of a smoke chamber would be a room with the door closed. If no intervening closed door exists between this and adjoining (communicating) rooms, the line denoting the barrier becomes less clear. The determining factor would be the depth of the wall section above the open archway or doorway, as follows:

- An archway or doorway that extends more than 18" down from the ceiling greatly delays smoke travel and is considered a boundary, just like a regular wall from floor to ceiling. A doorway that extends more than 4", but less than 18", must be considered a smoke barrier, and

spacing of detectors must be reduced to ⅔ of the detectors' listed spacing distance on either side of the opening (*Figure 51*).
- Open grids above doors or walls that allow free flow of air and smoke are not considered barriers. To be considered an open grid, the opening must meet all of the requirements defined and demonstrated in *Figure 52*.
- Smoke doors, which are kept open by hold-and-release devices and meet all of the standards applicable thereto, may also be considered boundaries of a smoke chamber.
- If the space between the top of a low wall and the ceiling is less than 18", the wall is treated as if it extends to the ceiling and is a barrier. Smoke will still be able to travel to the other side of the wall, but it may be substantially delayed. This will delay notification. In this case, detector spacing must be reduced to ⅔ of the listed spacing on either side of the wall. If the space between the top of a low wall and the ceiling is 18" or more, the wall is not considered to substantially affect the smoke travel.

Wireless Smoke Detection System

The components that form a wireless smoke detection system are shown here. This type of system can be used in situations where the building design or installation costs make hard wiring of the detectors impractical or too expensive. The heart of this system is the translation unit, called a gateway. It can communicate with up to four remote receiver units. The receivers can monitor radio frequency signals from up to 80 wireless smoke detectors. Each receiver unit transmits the status of the related wireless detectors via communication wiring to the translator. The translator then communicates this status to an intelligent FACP via a signaling line circuit loop. (Note that some jurisdictions do not permit the use of these systems.)

26405-14_SA09.EPS

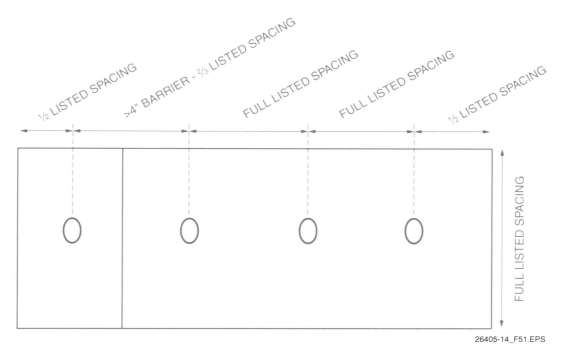

Figure 51 Reduced spacing required for a barrier.

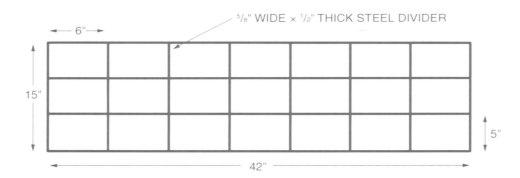

GRID OPENINGS MUST BE AT LEAST ¹/₄" IN THE LEAST DIMENSION.
OPENINGS ARE 5³/₈" (6" MINUS ⁵/₈") × 4³/₈" (5" MINUS ⁵/₈").
REQUIREMENT MET.

THE THICKNESS OF THE MATERIAL DOES NOT EXCEED THE LEAST DIMENSION.
OPENINGS ARE 5³/₈" (6" MINUS ⁵/₈") × 4³/₈" (5" MINUS ⁵/₈"). THICKNESS IS ¹/₂".
REQUIREMENT MET.

THE OPENINGS CONSTITUTE AT LEAST 70% OF THE AREA OF THE PERFORATED MATERIAL.
OPENINGS ARE 5³/₈" (6" MINUS ⁵/₈") × 4³/₈" (5" MINUS ⁵/₈")
5³/₈" × 4³/₈" = 23¹/₂ sq. in.

21 OPENINGS × 23¹/₂ sq. in.   =   TOTAL OPENING = 493¹/₂ sq. in.
15" × 42" = 630 sq. in.   630 sq. in. × 0.70 = 441 sq. in.
REQUIREMENT MET.

26405-14_F52.EPS

Figure 52 Detailed grid definition.

11.3.2 Smoke Spread and Stratification Phenomena

In a fire, smoke and heat rise in a plume toward the ceiling because they are lighter than the more dense, surrounding cooler air. In an area with a relatively low, flat, smooth ceiling, the smoke and heat quickly spread across the entire ceiling, triggering smoke or heat detectors. When the ceiling is irregular, smoke and heated air will tend to collect, perhaps stratifying near a peak or collecting in the bays of a beamed or joist ceiling.

In the case of a beamed or joist ceiling, the smoke and heated air will fill the nearest bays and begin to overflow to adjacent bays (*Figure 53*). The process continues until smoke and heat reach either a smoke or heat detector in sufficient quantity to cause activation. Because each bay must fill before it overflows, a substantial amount of time may pass before a detector placed at its maximum listed spacing activates. To reduce this time, the spacing for the detectors is reduced.

When smoke must rise a long distance, it tends to cool off and become denser. As its density becomes equal to that of the air around it, it stratifies, or stops rising (*Figure 54*). As a fire grows, additional heat is added, and the smoke will eventually rise to the ceiling. However, a great deal of time may be lost, the fire will be much larger, and a large quantity of toxic gases will be present at the floor level. Due to this delay, alternating detectors are lowered at least 3' (*Figure 55*). The science of stratification is very complex and requires a fire protection engineer's evaluation to determine if stratification is a factor and, if so, to determine how much to lower the detectors to compensate for this condition.

Some conditions that cause stratification include:

- Uninsulated roofs that are heated by the sun, creating a heated air thermal block
- Roofs that are cooled by low outside temperatures, cooling the gases before they reach a detector
- HVAC systems that produce a hot layer of ceiling air
- Ambient air that is at the same temperature as the fire gases and smoke

There are no clear, set rules regarding which environments will and will not be susceptible to stratification. The factors and variables involved are beyond the scope of this course. For further reading on the subject, contact the NFPA reference department for a bibliography. Note that only alternating detectors are suspended for stratification. Some fires can result in stratification caused

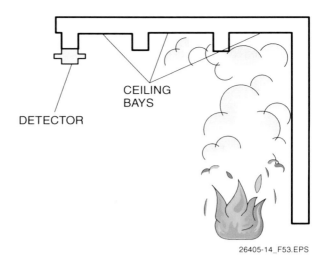

26405-14_F53.EPS

Figure 53 Smoke spread across a beamed or joist ceiling.

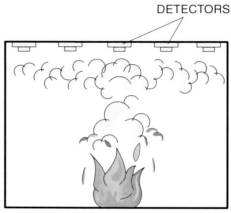

26405-14_F54.EPS

Figure 54 Smoke stratification.

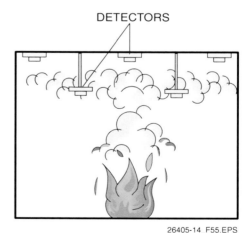

26405-14_F55.EPS

Figure 55 Smoke stratification countermeasure.

by superheated air and gases reaching the ceiling prior to the smoke, causing a barrier that holds the smoke below the detectors. Smoke detectors with integral heat detectors should be effective in such cases, without the need to suspend alternating detectors below the ceiling.

Stratification can also occur in high-rise buildings because of the stack effect principle. This occurs when smoke rises until it reaches a neutral pressure level, where it begins to stratify (*Figure 56*). Stack effect can move smoke from the source to distant parts of the building. Penetrations left unsealed can be a major contributor in such smoke migration. Stack effect factors include:

- Building height
- Air tightness
- Air leakage between floors
- Interior/exterior temperature differential
- Vertical openings
- Wind force and direction

Some factors that influence stack effect are variable, such as the weather conditions and which interior and exterior doors might be open. For this reason, the neutral pressure level may not be the same in a given building at different times. One item is very clear concerning stratification: penetrations made in smoke barriers, particularly those in vertical openings such as shafts, must be sealed.

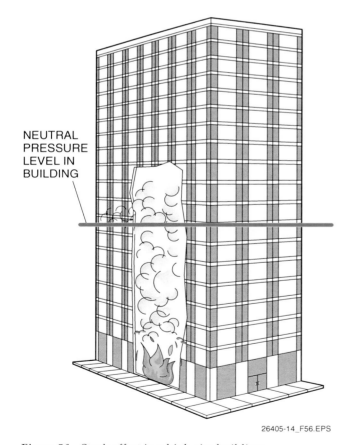

NEUTRAL PRESSURE LEVEL IN BUILDING

26405-14_F56.EPS

Figure 56   Stack effect in a high-rise building.

11.4.0 General Precautions for Detector Installation

The following are general precautions for detector installation:

- *Recessed mounting* – Detectors must not be recess-mounted unless specially listed for recessed mounting.
- *Air diffusers* – Air movement can have a number of undesirable effects on detectors. The introduction of air from outside the fire area can dilute the smoke, which delays activation. The air movement may create a barrier, which delays or prevents smoke from reaching the detector chamber. Smoke detectors are listed for a specific range of air velocity. This means that they are expected to function with air (and smoke) moving through the detection chamber at any speed within the listed range. Detectors may not function properly if air movement is above or below the listed velocity. Air movement can be measured with an instrument called a velocimeter. Unless the airflow exceeds the listed velocity, spot-type smoke detectors must be a minimum of 3' from air diffusers.
- *Problem locations and sources of false alarms* – Detectors, especially smoke detectors, can register false alarms as a result of the following:
  - *Electrical interference:* Keep detector locations away from fluorescent lights and radio transmitters, including cellular phones. Electrical noise or radio frequencies radiated from these devices can cause false alarms.
  - *Heating equipment:* High temperatures, dust accumulation, improper exhaust, and incomplete combustion from heating equipment are problems for detectors.
  - *Engine exhaust:* Exhaust from engine-powered forklifts, vehicles, and generators is a

potential problem for ionization-type smoke detectors.

- *Solvent and chemical fumes:* Cleaning solvents and adhesives are a problem for ionization detectors.
- *Other gases and fumes:* Fumes from machining operations, paint spraying, industrial or battery gases, curing ovens/dryers, cooking equipment, sawing/drilling/grinding operations, and welding/cutting operations cause problems for smoke detectors.
- *Extreme temperatures:* Avoid very hot or cold environments for smoke detector locations. Temperatures below 32°F (0°C) can cause false alarms, and temperatures above 120°F (48.8°C) can prevent proper smoke detector operation. Extreme temperatures affect beam, ionization, and photoelectric detectors.
- *Dampness or humidity:* Smoke detectors must be located in areas where the humidity is less than 93%. In ionization detectors, dampness and high humidity can cause tiny water droplets to condense inside the sensing chamber, making it overly sensitive and causing false alarms. In photoelectric detectors, humidity can cause light refraction and loss of current flow, either of which can lead to false alarms. Common sources of moisture to avoid include slop sinks, steam tables, showers, water spray operations, humidifiers, and live steam sources.
- *Lightning:* Nearby lightning can cause electrical damage to a fire alarm system. It may also cause electrical noise or spikes to be induced in the alarm system wiring or detectors, resulting in false alarms. Surge arrestors installed in the system's primary power supply to protect the system, in conjunction with alarm verification, can reduce the chance of system damage and false alarms.
- *Dusty or dirty environments:* Dust and dirt can accumulate on a smoke detector's sensing chamber, making it overly sensitive. Avoid areas where fumigants, fog, dust, or mist-producing materials are consistently used.
- *Outdoor locations:* Dust, air currents, and humidity typically affect outdoor structures, including sheds, barns, stables, and other open structures. This makes outdoor structures unsuitable for smoke detectors.
- *Insect-infested areas:* Insects in a smoke detector can cause a false alarm. Good bug screens

Troubleshooting

Troubleshooting older systems requires that you trace and check each detector in order to isolate the source of an alarm. The fire alarm panels of newer systems, such as the one shown here, have fault isolation messages and LCD display panels to help pinpoint which detector has alarmed or has a problem.

26405-14_SA10.EPS

on a detector can prevent most adult insects from entering the detector. However, newly hatched insects may still be able to enter. An insecticide strip next to the detector may help solve the problem, but it may also cause false alarms because of fumes. Check with the manufacturer for the use of an approved strip. Ionization detectors are less prone to false alarms from insects.

- *Construction:* Smoke detectors must not be installed until after construction cleanup unless required by the AHJ for protection during construction. Detectors that are installed prior to construction cleanup must be cleaned or replaced. Prior to 1993, the code permitted covering smoke detectors to protect them from dirt, dust, or paint mist. However, the covers did not work very well and resulted in clogged, oversensitive, and damaged detectors.

11.5.0 Spot Detector Installations on Flat, Smooth Ceilings

Flat, smooth ceilings are defined as ceilings that have a slope equal to or less than 1.5" per foot (1' to 8') and do not have open joists or beams.

11.5.1 Conventional Installation Method

The following applies to conventional spot detector installation on flat, smooth ceilings:

- When a smoke detector manufacturer's specifications do not specify a particular spacing, a 30' spacing guide may be applied.

> **NOTE**
>
> This section assumes detectors listed for 30' spacing mounted on a smooth and flat ceiling of less than 10' in height, where 15' is one-half the listed spacing. Spacing is reduced as indicated in later sections when the **ceiling height** exceeds 10', or the ceiling is not smooth and flat as defined by the code.

- The distance between heat detectors must not exceed their listed spacing.
- There must be detectors located within one-half the listed spacing measured at right angles from all sidewalls.

In the following example, the detector locations for a simple room that is 30' × 60' long will be determined. The locations are determined by the intersection of columns and rows marked on a sketch of the ceiling:

- The first column is located by a line that is parallel to the end wall and not more than one-half the listed spacing from the end wall (*Figure 57*).
- The first row is located by a line parallel to the sidewall that is not more than one-half the listed spacing from that sidewall (*Figure 58*).
- The first detector is located at the intersection of the row and column lines.
- The second column is located by a line that is parallel to the opposite end wall and not more than one-half the listed spacing from the end wall.
- The second detector is located at the intersection of the row and second column lines, provided that the distance between the first and second detectors does not exceed the listed spacing (*Figure 59*). The only time that the full listed spacing of any detector is used is when measuring from one detector to the next.

11.6.0 Photoelectric Beam Smoke Detector Installations on Flat, Smooth Ceilings

Two configurations of beam detectors are used for open area installations. Beam smoke detectors can be installed as straight-line devices or as angled-beam devices that employ mirrors. When used as angled-beam devices, the beam length is reduced for the number of mirrors used, as specified in

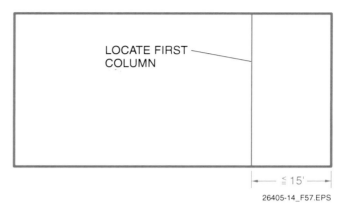

Figure 57　Locating the first column.

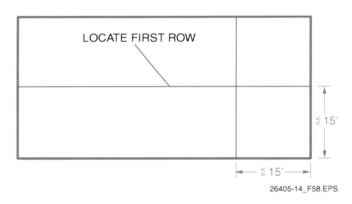

Figure 58　Locating the first row.

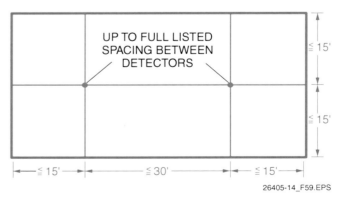

Figure 59　First and second detector locations.

NFPA 72®. When installing either type of beam smoke detector, always follow the manufacturer's instructions.

11.6.1 Straight-Line Beam Detector Installations

Photoelectric beam smoke detectors have basically two listings: one for width of coverage and one for the minimum/maximum length of coverage (beam length), as shown in *Figure 60*. The beam length is listed by the manufacturer and, in most cases, so is the width coverage spacing (S). When a manufacturer does not list the width coverage, the 60' guideline specified in *NFPA 72®* must be used as the width coverage. The distance of the beam from the ceiling should normally be between 4" and 12". However, NFPA allows a greater distance to compensate for stratification. Always make sure the beam does not cross any expansion joint or other point of slippage that could eventually cause misalignment of the beam. In large areas, parallel beam detectors may be installed, separated by no more than their listed spacing (S).

11.6.2 Angled-Beam Installations Using Mirrors

The use of mirrors to direct the light beam from the transmitter in a path other than a single straight line is permitted if the manufacturer's directions are followed and the mirror used is listed for the model of the beam detector being installed. The use of a mirror will require a reduction in the listed beam length of the detector. When using a mirror, the beam length is the distance between the transmitter and receiver (measured from the transmitter to the mirror, plus the distance from the mirror to the receiver). This distance is reduced to 66% of the manufacturer's listed beam length for a single mirror, and to 44% of the manufacturer's listed beam length for two mirrors. Maximum coverage for a two-mirror beam detector is shown in *Figure 61*. Width spacing (S) is defined as previously stated for single-beam and double-beam

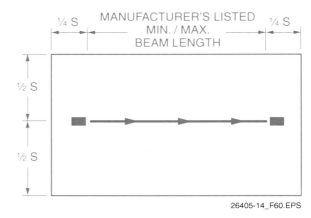

Figure 60 Maximum straight-line single-beam smoke detector coverage (ceiling view).

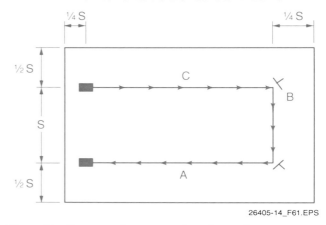

Figure 61 Coverage for a two-mirror beam detector installation.

installations. The planner should also allow for additional installation and service time when mirrors are used. As with beam detectors without mirrors, make sure that the light beam does not cross a building expansion joint.

11.7.0 Spot Detector Installations on Irregular Ceilings

Any ceiling that is not flat and smooth is considered irregular. For heat detectors, ceilings of any description that are over 10' above floor level are considered irregular. Heat detectors located on any ceiling over 10' above finished floor (AFF) level must have their spacing reduced below their listed spacing in accordance with *NFPA 72®*. This is because hot air is diluted by the surrounding cooler air as it rises, which reduces its temperature. Smoke detector spacing is not adjusted for high ceilings.

Irregular ceilings include sloped, solid joist, and beam ceilings:

- *Sloped ceilings* – A sloped ceiling is defined as a ceiling having a slope of more than 1.5" per foot (1' to 8'). Any smooth ceiling with a slope equal to or less than 1.5" per foot is considered a flat ceiling. Sloped ceilings are usually shed or peaked types.
  - *Shed:* A shed ceiling is defined as having the high point at one side with the slope extending toward the opposite side.
  - *Peaked:* A peaked ceiling is defined as sloping in two directions from its highest point. Peaked ceilings include domed or curved ceilings.
- *Solid joist or beam ceilings* – Solid joist or beam ceilings are defined as being spaced less than 3' center-to-center with a solid member extending down from the ceiling for a specified distance. For heat detectors, the solid member must extend more than 4" down from the ceiling. For smoke detectors, the solid member must extend down more than 8".

11.7.1 Shed Ceiling Detector Installation

Shed ceilings having a rise greater than 1.5"/1' run (1'/8') must have the first row of detectors (heat or smoke) located on the ceiling within 3' of the high side of the ceiling (measured horizontally from the sidewall).

Heat detectors must have their spacing reduced in areas with high ceilings (over 10'). Smoke detector spacing is not adjusted due to the ceiling height.

For a roof slope of less than 30°, determined as shown in *Figure 62*, all heat detector spacing must be reduced based on the height at the peak. For a roof slope of greater than 30°, the average ceiling height must be used for all heat detectors other than those located in the peak. The average ceiling height can be determined by adding the high side-wall and low sidewall heights together and dividing by two. In the case of slopes greater than 30°, the spacing to the second row is measured from the first row and not the sidewall for shed ceilings. See *NFPA 72®* for heat detector spacing reduction based on peak or average ceiling heights.

Once you have determined that the ceiling you are working with is a shed ceiling, use the following guidelines for determining detector placement:

- Place the first row of detectors within 3' of the high sidewall (measured horizontally).
- Use the listed spacing (adjusted for the height of the heat detectors on ceilings over 10') for

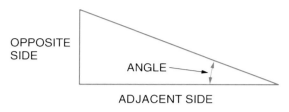

DETERMINING IF A SLOPE IS GREATER THAN 30°

OPPOSITE SIDE
ANGLE
ADJACENT SIDE

If the opposite side divided by the adjacent side is > 0.5774, the angle is > 30°.

The tangent of the smallest angle of a right triangle equals the opposite side divided by the adjacent side.

The sum of all the angles of a triangle equals 180°.

The tangent of a 30° angle is 0.5774.

26405-14_F62.EPS

Figure 62 Determining degree of slope.

each additional row of detectors (measured from the detector location, not the sidewall).
- For heat detectors on ceilings over 10' high, adjust the spacing per *NFPA 72®*.
- If the slope is less than 30°, all heat detector spacing is based on the peak height. If the slope is more than 30°, peak heat detectors are based on the peak height, and all other detectors are based on the average height.
- Additional columns of detectors that run at right angles to the slope of the ceiling do not need to be within 3' of the end walls.
- Smoke detector spacing is not adjusted for ceiling height.

To calculate the average ceiling height, use one of the following methods:

Method 1:

Step 1 Subtract the height of the low ceiling from the height of the high ceiling.

Step 2 Divide the above result by 2.

Step 3 Subtract the above result from the height of the high ceiling.

Method 2:

Step 1 Add the height of the low ceiling to the height of the high ceiling.

Step 2 Divide the above result by 2.

11.7.2 Peaked Ceiling Detector Installation

A ceiling must be sloped as defined by the code in order to be considered a peaked ceiling. It must slope in more than one direction from its highest point. Because domed and curved ceilings do not have clean 90° lines to clearly show that the ceiling slopes in more than one direction, they should be viewed from an imaginary vertical center line. From this perspective, it can be seen if the ceiling slopes in more than one direction.

- The first row of detectors (heat or smoke) on a peaked ceiling should be located within 3' of the peak, measured horizontally. The detectors may be located alternately on either side of the peak, if desired.
- Regardless of where in the 3' space on either side of the peak the detector is located, the measurement for the next row of heat detectors is taken from the peak (measured horizontally, not from the heat detector located near the peak). This is different from the way detector location is determined for shed ceilings.
- Heat detectors must have their spacing reduced in areas with ceilings over 10'. Smoke detector spacing is not adjusted for ceiling height.
- As with sloped ceilings, use of the peak height for just the row of heat detectors at the peak, or for all the ceiling heat detectors, depends on whether or not the ceiling is sloped more than 30°. Use the peak height to adjust heat detector spacing of all the room's heat detectors when the slope of the ceiling is less than 30°. When the slope is greater than 30°, use the peak height for only the heat detectors located within 3' of the peak. Use the average height for all other heat detectors in the room.

Smoke detectors should be located on a peaked ceiling such that:

- A row of detectors is within 3' of the peak on either side or alternated from side to side.
- The next row of detectors is within the listed spacing of the peak measured horizontally. Do not measure from the detectors within 3' of the peak. It does not matter where within 3' of the peak the first row is; the next row on each side of the peak is installed within the listed spacing of the detector from the peak.
- Additional rows of detectors are installed using the full listed spacing of the detector.
- The sidewall must be within one-half the listed spacing of the last row of detectors.
- Columns of detectors are installed in the depth dimension of the above diagrams using the full listed spacing (the same as on a smooth, flat ceiling).
- There is no need to reduce smoke detector spacing for ceilings over 10'.

11.7.3 Solid Joist Ceiling Detector Installation

Solid joists are defined for heat detectors as being solid members that are spaced less than 3' center-to-center and that extend down from the ceiling more than 4". Solid joists are defined for smoke detectors as being solid members that are spaced less than 3' center-to-center and that extend down from the ceiling more than 8".

Heat detector spacing at right angles to the solid joist is reduced by 50%. In the direction running parallel to the joists, standard spacing principles are applied. If the ceiling height exceeds 10', spacing is adjusted for the high ceiling in addition to the solid joist as defined in *NFPA 72®*.

Smoke detector spacing at right angles to the solid joists is reduced by one-half the listed spacing for joists 1' or less in depth and ceilings 12' or less in height. In the direction running parallel to the joists, standard spacing principles are applied. If the ceiling height is over 12' or the depth of the joist exceeds 1', spot-type detectors must be located on the ceiling in every pocket. Additional reductions for sloped joist ceilings also apply as defined in *NFPA 72®*.

11.7.4 Beamed Ceiling Detector Installation

Beamed ceilings consist of solid structural or solid nonstructural members projecting down from the ceiling surface more than 4" and spaced more than 3' apart center-to-center as defined by *NFPA 72®*.

- *Heat detector installation:*
  - If the beams project more than 4" below the ceiling, the spacing of spot-type heat detectors at right angles to the direction of beam travel must not be more than two-thirds the smooth ceiling spacing.
  - If the beams project more than 12" below the ceiling and are more than 8' on center, each bay formed by the beams must be treated as a separate area.
  - Reductions of heat detector spacing in accordance with *NFPA 72®* are also required for ceilings over 10' in height.
- *Smoke detector installation:*
  - For smoke detectors, if beams are 4" to 12" in depth with an AFF level of 12' or less, the spacing of spot-type detectors in the direction perpendicular to the beams must be reduced 50%, and the detectors may be mounted on the bottoms of the beams.
  - If beams are greater than 12" in depth with an AFF level of more than 12', each beam bay must contain its own detectors. There are additional rules for sloped beam ceilings as defined in *NFPA 72®*.

- The spacing of projected light beam detectors that run perpendicular to the ceiling beams need not be reduced. However, if the projected light beams are run parallel to the ceiling beams, the spacing must be reduced per *NFPA 72®*.

11.8.0 Notification Appliance Installation

There are several types of notification appliances:

- Audible devices such as bells, horns, chimes, speakers, sirens, and mini-sounders
- Visual devices such as strobe lights
- Tactile (sense of touch) devices such as bed shakers and special sprinkling devices
- Olfactory (sense of smell) devices, which are also accepted by the code

The appropriate devices for the occupancy must be determined. All devices must be listed for the purposes for which they are used. For example, *UL 1480* speakers are used for fire protective signaling systems, and *UL 1638* visual signaling appliances are used for private mode emergency and general utility signaling. While *NFPA 72®* recognizes tactile and olfactory devices, it does not specify installation requirements. If an occupant requires one of these types of notification, use equipment listed for the purpose and follow the manufacturer's instructions. As always, consult the local AHJ.

11.8.1 Notification Device Installation

The following guidelines apply to the installation of notification devices:

- Ensure that notification devices are wired using the applicable circuit style (Class A or B).
- In Class B circuits, the panel supervises the wire using an end-of-line device. Electrically, the EOL device must be at the end of the indicating circuit. Examples of EOL devices are resistors to limit current, diodes for polarity, and capacitors for filtering.
- It is extremely important that any polarized devices be installed correctly. The polarization of the leads or terminals of these devices are marked on the device or are noted in the manufacturer's installation data. If the leads of a polarized notification device are reversed, the panel will not detect the problem, and the device will not activate. Moreover, the device will act as an EOL device, preventing the panel from detecting breaks between the device and the actual EOL device. Because it is very easy to accidentally wire a device backwards, testing every device is extremely important. The gen-

eral alarm should be activated and every notification device checked for proper operation.
- Sidewall-mounted audible notification devices must be mounted at least 90" AFF level or at least 6" below the finished ceiling. Ceiling-mounting and recessed appliances are permitted.
- Visible notification appliances must be mounted at the minimum heights of 80" to 96" AFF (*NFPA 72®*) or, for ADA requirements, either 80" AFF or 6" below the ceiling. In any case, the device must be within 16' of the pillow in a sleeping area. Combination audible/visible appliances must follow the requirements for visible appliances. Non-coded visible appliances should be installed in all areas where required by *NFPA 101®* or by the local AHJ. Consult ADA codes for the required illumination levels in sleeping areas.
- *Visual notification appliance spacing in corridors–* *Table 5* provides spacing requirements for corridors less than 20' wide. For corridors and rooms greater than 20', refer to *Tables 6* and *7*. In corridor applications, visible appliances must be rated at not less than 15 candelas (cd). Per *NFPA 72®*, visual appliances must be located no more than 15' from the end of the corridor with a separation of no more than 100' between appliances. Where there is an interruption of the concentrated viewing path, such as a fire door or elevation change, the area is to be considered as a separate corridor.
- *Visual notification appliance spacing in other applications* – The light source color for visual appliances must be clear or nominal white and must not exceed 1,000cd (*NFPA 72®*). In addition, special considerations apply when more than one visual appliance is installed in a room or corridor. *NFPA 72®* specifies that the separation between appliances must not exceed 100'. Visible notification appliances must be installed

Table 5 Visual Notification Devices Required for Corridors not Exceeding 20'

Corridor Length (in ft.)	Minimum Number of 15cd Appliances Required
0–30	1
31–130	2
131–230	3
231–330	4
331–430	5
431–530	6

Table 6 Room Spacing for Wall-Mounted Visual Notification Appliances

Maximum Room Size (in ft.)	Minimum Required Light Output in Candelas (cd)		
	One Light per Room	Two Lights per Room*	Four Lights per Room**
20 × 20	15	Not allowable	Not allowable
30 × 30	30	15	Not allowable
40 × 40	60	30	Not allowable
50 × 50	95	60	Not allowable
60 × 60	135	95	Not allowable
70 × 70	185	95	Not allowable
80 × 80	240	135	60
90 × 90	305	185	95
0 × 100	375	240	95
110 × 110	455	240	135
120 × 120	540	305	135
130 × 130	635	375	185

* Locate on opposite walls
** One light per wall

Table 7 Room Spacing for Ceiling-Mounted Visual Notification Appliances

Maximum Room Size (in ft.)	Maximum Ceiling Height (in ft.)*	Minimum Required Light Output for One Light (cd)**
20 × 20	10	15
30 × 30	10	30
40 × 40	10	60
50 × 50	10	95
20 × 20	20	30
30 × 30	20	45
40 × 40	20	80
50 × 50	20	95
20 × 20	30	55
30 × 30	30	75
40 × 40	30	115
50 × 50	30	150

* Where ceiling heights exceed 30', visible signaling appliances must be suspended at or below 30' or wall mounted in accordance with *NFPA 72*®.
** This table is based on locating the visible signaling appliance at the center of the room. Where it is not located at the center of the room, the effective intensity (cd) must be determined by doubling the distance from the appliance to the farthest wall to obtain the maximum room size.

in accordance with *Tables 6* and *7*, using one of the following:
- A single visible notification appliance
- Two visible notification appliances located on opposite walls
- More than two appliances for rooms 80' × 80' or larger (must be spaced a minimum of 55' from each other)

11.9.0 Fire Alarm Control Panel Installation Guidelines

The guidelines for installing a fire alarm control panel are as follows:

• When not located in an area that is continuously occupied, all fire alarm control equipment must be protected by a smoke detector, as shown in *Figure 63*. If the smoke detector is not designed to work properly in that environment, a heat detector must be used. It is not necessary to protect the entire space or room.

• Detector spacing must be adjusted if the ceiling over the control equipment is irregular in one or more respects. This is considered protection against a specific hazard under *NFPA 72®* and does not require the entire chamber (room) containing the control equipment to be protected under the 0.7 rule.

• A means of silencing audible notification appliances from an FACP must be protected against unauthorized use. Most FACPs are located inside locked metal cabinets as shipped from the manufacturer. If the silencing switch is

On Site

UL 1971 Listing

A *UL 1971* listing for a visual notification appliance indicates that it meets the ADA hearing-impaired requirements.

key-actuated or is locked within the cabinet, this provision should be considered satisfied. If the silencing means is within a room that is restricted to authorized use only, no additional measures should be required. The NFPA codes do not clearly define nor specify unauthorized or authorized use.

• FACP connections to the primary light and power circuits must be on a dedicated branch circuit with overcurrent protection rated at 20A or less. Any connections to the primary power circuit on the premises connected after the distribution panel (circuit breaker box) must be directly related to the fire alarm system. No other use is permitted. This requirement does not necessitate a direct tap into the power circuit ahead of the distribution panel, although connecting ahead of the main disconnect is acceptable with listed service equipment.

• The power disconnection means (circuit breaker) must be clearly identified as a fire alarm circuit control.

12.0.0 FIRE ALARM-RELATED SYSTEMS AND INSTALLATION GUIDELINES

This section discusses various fire alarm-related systems, as well as the installation guidelines for each system.

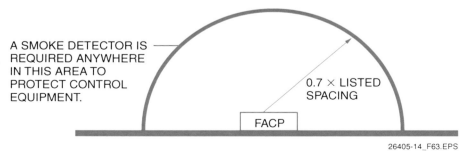

A smoke detector protecting control equipment must meet *NFPA 72* spacing and placement standards, but the entire space (room) containing the FACP need not be protected.

A SMOKE DETECTOR IS REQUIRED ANYWHERE IN THIS AREA TO PROTECT CONTROL EQUIPMENT.

0.7 × LISTED SPACING

FACP

26405-14_F63.EPS

Figure 63   Protection of an FACP.

12.1.0 Ancillary Control Relay Installation Guidelines

Ancillary functions, commonly called auxiliary functions, include such controls as elevator capture (recall), elevator shaft pressurization, HVAC system shutdown, stairwell pressurization, smoke management systems, emergency lighting, door unlocking, door hold-open device control, and building music system shutoff. For example, sound systems are commonly powered down by the fire alarm system so that the evacuation signal may be heard. In the normal state (*Figure 64*), an energized relay completes the power circuit to the device being controlled.

> **NOTE**
>
> The circuit from the relay to the background music system is a remote control signaling circuit (see *NEC Article 725*). If this circuit is both powered by and controlled by the fire alarm, the circuit is a fire alarm circuit (see *NEC Article 760*).

12.2.0 Duct Smoke Detectors

Duct smoke detectors are not simply conventional detectors applied to HVAC systems. Conventional smoke detectors are listed for open area protection (OAP) under *UL 268*. Duct smoke detectors are normally listed for a slightly higher velocity of air movement and are tested under *UL 268A*. The primary function of a duct smoke detector is to turn off the HVAC system. This prevents the system from spreading smoke rapidly throughout the building and stops the system from providing a forced supply of oxygen to the fire. Duct smoke detectors are not intended primarily for early warning and notification. Relevant fire alarm-related provisions pertaining to HVAC systems can be found in *NFPA 90A* and *B*.

NFPA 90A covers the following types of buildings and systems:

- HVAC systems serving spaces over 25,000 cubic feet in volume
- Buildings of Type III, IV, or V construction that are over three stories in height (see *NFPA 220*)
- Buildings that are not covered by other applicable standards
- Buildings that serve occupants or processes not covered by other applicable standards

NFPA 90B covers the following types of systems:

- HVAC systems that service one- or two-family dwellings
- HVAC systems that service spaces not exceeding 25,000 cubic feet in volume

> **NOTE**
>
> No duct smoke detector requirements are found under *NFPA Standard 90B*.

12.2.1 Duct Detector Location

When determining the location of duct detectors prior to installation, use the following guidelines:

- In HVAC units over 2,000 cubic feet per minute (cfm), duct detectors must be installed on the supply side.
- Duct detectors must be located downstream of any air filters and upstream of any branch connection in the air supply.
- Duct detectors should be located upstream of any in-duct heating element.

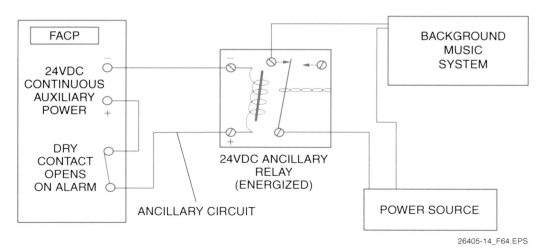

Figure 64 Music system control in normal state.

- Duct detectors must be installed at each story prior to the connection to a common return and prior to any recirculation of fresh air inlet in the air return of systems over 15,000 cfm serving more than one story.
- Return air system smoke detectors are not required when the entire space served by the HVAC system is protected by a system of automatic smoke detectors and when the HVAC is shut down upon activation of any of the smoke detectors.

12.2.2 Conversion Approximations

The approximations given in *Table 8* are useful when the protected-premises personnel do not know the cfm rating of an air-handling unit, but do know either the tonnage or British thermal unit (Btu) rating. Additionally, the cfm rating may not always appear on air handling unit (AHU) nameplates or in building specifications.

12.2.3 More Than One AHU Serving an Area

When more than one air handling unit (AHU) is used to supply air to a common space, and the return air is drawn from this common space, the total capacity of all units must be used in determining the size of the HVAC system (*Figure 65*). This formal interpretation makes clear the

fact that interconnected air handling units should be viewed as a system, as opposed to treating each AHU individually. When multiple AHUs serve a common space, the physical location of each AHU, relative to others interconnected to the same space, is irrelevant to the application of the formal interpretation.

The common space being served by multiple air handling units need not be contiguous (connected) for the formal interpretation to apply. In *Figure 66*, AHUs #3 and #4 serve common space and must be added together for consideration (1,100 + 1,100 = 2,200). Because 2,200 is greater than 2,000, duct smoke detectors are required on both. The same is true for AHU #1 and #2.

Duct smoke detectors may not be used as substitutes where open area detectors are required. This is because the HVAC unit may not be running when a fire occurs. Even if the fan is always on, the HVAC is not a listed fire alarm device.

Duct smoke detectors must automatically stop their respective fans upon detecting smoke. It is also acceptable for the fire alarm control panel to stop the fan(s) upon activation of the duct detector. However, fans that are part of an engineered smoke control or management system are an exception and are not always shut down in all cases when smoke is detected.

12.2.4 Duct Smoke Detector Installation and Connections

When a fire alarm system is installed in a building, all duct smoke detectors in that building must be connected to the fire alarm system as either initiating devices or as supervisory devices. The code does not require the installation of a building fire alarm system. It does require that the duct smoke detectors be connected to the building fire alarm, if one exists. Duct smoke detectors, properly listed and installed, will accomplish their intended function when connected as initiating or supervisory devices.

Table 8 Conversions

Capacity Rating	CFM
1 ton	400
12,000 Btus	400
5 tons	2,000
60,000 Btus	2,000
37.5 tons	15,000
450,000 Btus	15,000

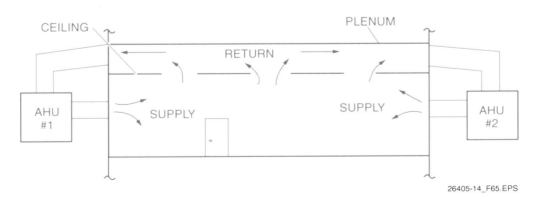

Figure 65 Non-ducted multiple AHU system.

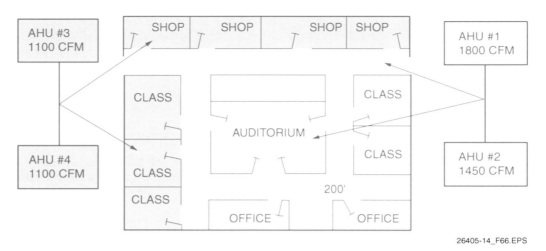

26405-14_F66.EPS

Figure 66 Example in which all AHUs require detectors on the supply side.

Remote Duct Smoke Indicator

This duct smoke indicator combines a horn with a key-activated test and reset function. Green, yellow, and red LEDs provide a visual indication of system power, a trouble condition, and an alarm, respectively. This unit is also equipped with an optional strobe and smoke lens that provides an enhanced visual indication of an alarm.

26405-14_SA11.EPS

When the building is not equipped with a fire alarm system, visual and audible alarm and trouble signal indicators (*Figure 67*) must be installed in a normally occupied area. Smoke detectors whose sole function is stopping fans do not require standby power.

AHUs often come with factory-installed duct detectors. If they are installed in a building with a fire alarm system, they must be connected to the fire alarm system, and the detector may have to have standby power. Consult with the AHJ concerning how many duct detectors can be placed in a zone. If it cannot be determined which detector activated after the power failure, the alarm indicator on the detector may be considered an additional function by the local AHJ.

26405-14_F67.EPS

Figure 67 Typical remote duct indicator.

Duct detectors for fresh air or return air ducts should be located six to ten duct widths from any openings, deflectors, sharp bends, or branch connections. This is necessary to obtain a representative air sample and to reduce the effects of stratification and dead air space.

12.3.0 Elevator Recall

Elevator recall is intended to route an elevator car to a non-fire floor, open its doors, and then put the car out of normal service until reset (Phase 1 recall). It is usually assumed that the recall floor will be the level of primary exit discharge or a grade level exit. In the event that the primary level of exit discharge is the floor where the fire has been detected, the system should route the car to a predetermined alternate floor, open its doors, and go out of normal service (Phase 2 recall). The operation of the elevators must be in accordance with *ANSI A17.1*.

Phase 1 recall (*Figure 68*) is the method of recalling the elevator to the floor that provides the highest probability of safe evacuation (as determined by the AHJ) in the event of a fire emergency.

Phase 2 recall (*Figure 69*) is the method of recalling the elevator to the floor that provides the next highest probability of safe evacuation (as determined by the AHJ) in the event of a fire emergency. Phase 2 recall is typically activated by the smoke detector in the lobby where the elevator would normally report during a Phase 1 recall.

Hoistway and elevator machine room smoke detectors must report an alarm condition to a building fire alarm control panel as well as an indicator light in the elevator car itself. However,

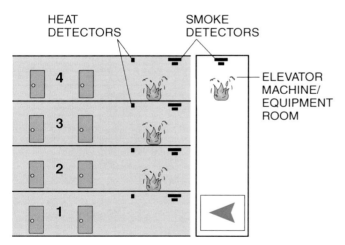

PHASE 1 RECALL:
Activation of any of these detectors sends all cars to the designated level.

26405-14_F68.EPS

Figure 68   Phase 1 recall.

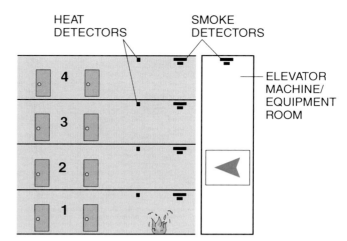

PHASE 2 RECALL:
Activation of the designated level detector sends all cars to the alternate floor.

26405-14_F69.EPS

Figure 69   Phase 2 recall.

notification devices may not be required to be activated if the control panel is in a constantly attended location. In facilities without a building fire alarm system, these smoke detectors must be connected to a dedicated fire alarm system control unit that must be designated as an Elevator Recall Control and Supervisory Panel.

Elevator recall must be initiated only by elevator lobby, hoistway, and machine room smoke detectors. Activation of manual pull stations, heat detectors, duct detectors, and any smoke detector not mentioned previously must not initiate elevator recall. In many systems, it would be inappropriate for the fire alarm control panel to initiate elevator recall. An exception is any system where the control function is selectable or programmable and the configuration limits the recall function to the specified detector activation only.

Caution should be used when using two-wire smoke detectors to recall elevators. Each elevator lobby, elevator hoistway, and elevator machine room smoke detector must be capable of initiating elevator recall when all other devices on the same initiating device circuit (IDC) have been manually or automatically placed in the alarm condition.

Unless the area encompassing the elevator lobby is a part of a chamber being protected by smoke detectors, such as the continuation of a corridor, the smoke detectors serving the elevator lobby (*Figure 70*) may be applied to protect against a specific hazard per *NFPA 72®*. The elevator lobby, hoistway, or associated machine room detectors may also be used to activate emergency control functions as permitted by *NFPA 101®*.

Where ambient conditions prohibit installation of automatic smoke detection, other appropriate

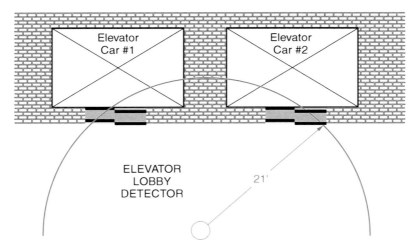

Assumes a 30' listed detector on a smooth and flat ceiling.
Target protection should be the extreme edges of the elevator door opening.

26405-14_F70.EPS

Figure 70 Elevator lobby detector.

automatic fire detection must be permitted. For example, a heat detector may protect an elevator door that opens to a high-dust or low-temperature area. Always refer to local codes.

12.4.0 Special Door Locking Arrangements

No lock, padlock, hasp, bar, chain, or other device intended to prevent free egress may be used at any door on which panic hardware or fire exit hardware is required (*NFPA 101®*). Note that these requirements apply only to exiting from the building. There is no restriction on locking doors against unrestricted entry from the exterior.

12.4.1 Stair Enclosure Doors

Upon activation of the fire alarm, stair enclosure doors must unlock to permit egress. They must also unlock to permit reentry into the building floors. This provision applies to buildings four stories and taller. The purpose of this requirement is to provide for an escape from fire and smoke entering a lower level of the stair enclosure and blocking safe egress from the stair enclosure. See *NFPA 101®* for additional details and exceptions.

12.5.0 Suppression System Supervision

The following section discusses both dry and wet chemical extinguishing systems.

12.5.1 Dry Chemical Extinguishing Systems

Dry chemical extinguishing systems must be connected to the building fire alarm system if a fire alarm system exists in the structure. The extinguishing system must be connected as an initiating device. The standard (*NFPA 17, Dry Chemical Extinguishing Systems*) does not require the installation of a building fire alarm system if one was not required elsewhere. Also see *NFPA 12* (*Carbon Dioxide Extinguishing Systems*), *NFPA 12A* (*Halon 1301 Fire Extinguishing Systems*), and *NFPA 2001* (*Clean Agent Fire Extinguishing Systems*).

12.5.2 Wet Chemical Extinguishing Systems

Wet chemical extinguishing systems must be connected to the building fire alarm system if a fire alarm system exists in the structure. As with dry chemical extinguishing systems, the standard (*NFPA 17A, Wet Chemical Extinguishing Systems*) does not require the installation of a building fire alarm system. Also, see *NFPA 16* (*Foam-Water Sprinkler and Foam-Water Spray Systems*).

12.6.0 Supervision of Suppression Systems

Each of the occupancy chapters of *NFPA 101®* will specify the extinguishing requirements for that occupancy. The code will specify either an approved automatic sprinkler or an approved supervised automatic sprinkler system if a sprinkler system is required.

12.6.1 Supervised Automatic Sprinkler

Where a supervised automatic sprinkler is required, various sections of *NFPA 101®* are applicable. These code sections begin with the phrase,

"Where required by another section of this code." Use of the word *supervised* in the extinguishing requirement section of each occupancy chapter is the method used to implement these two provisions. Sprinkler systems do not automatically require electronic supervision. Supervised automatic sprinkler systems require a distinct supervisory signal. This signal indicates a condition that would impair the proper operation of the sprinkler system, at a location constantly attended by qualified personnel or at an approved remote monitoring facility. Water flow alarms from a supervised automatic sprinkler system must be transmitted to an approved monitoring station. When the supervised automatic sprinkler supervisory signal terminates on the protected premises in areas that are constantly attended by trained personnel, the supervisory signal is not required to be transmitted to a monitoring facility. In such cases, only alarm and trouble signals need to be transmitted.

The following are some of the sprinkler elements that are required to be supervised where applicable:

• Water supply control valves
• Fire pump power (including phase monitoring)
• Fire pump running
• Water tank levels
• Water tank temperatures
• Tank pressures
• Air pressure of dry-pipe systems

A sprinkler flow alarm must be initiated within 90 seconds of the flow of water equal to or greater than the flow from the sprinkler head, or from the smallest orifice (opening) size in the system. In actual field verification activities, the 90 seconds is measured from the time water begins to flow into the inspector's test drain, and *not* from the time the inspector's test valve is opened. The smallest orifice is the size of the opening at the smallest sprinkler head.

12.6.2 Sprinkler Systems and Manual Pull Stations

Sprinkler systems that initiate a fire alarm system by a water flow switch must include at least one manual station that is located where required by the AHJ. Manual stations required elsewhere in the codes or standards can be considered to meet this requirement. Some occupancy chapters of *NFPA 101®* will allow the sprinkler flow switch to substitute for manual stations at all the required exits. If such an option is utilized, this provision requires that at least one manual station be installed where acceptable to the AHJ.

12.6.3 Outside Screw and Yoke Control Valves and Tamper Switches

Each shutoff valve, also called an outside screw and yoke (OS&Y) control valve, must be supervised by a tamper switch. A distinctive signal must sound when the valve is moved from a fully open (off-normal) position (within two revolutions of the hand wheel, or when the stem has moved one-fifth from its normal open position). A common verification practice is to mark the 12:00 position on the valve in its normal (open) position, then rotate the hand wheel twice and stop at the 12:00 position on the second pass. The supervisory signal must be initiated by the time the wheel reaches the end of the second pass.

Water flow and supervisory devices, in addition to their circuits, must be installed such that no unauthorized person may tamper with them, open them, remove them, or disconnect them without initiating a signal. Publicly accessible junction boxes must have tamper-resistant screws or tamper-alarm switches. Most water flow switch and valve tamper switch housings come from the manufacturer with tamper-resistant screws (hex or allen head), and they may be configured to signal when the housing cover is removed. If the device, circuit, or junction box requiring protection is in an area that is not accessible to unauthorized personnel, no additional protective measures should be required. Simply sealing, locking, or removing the handle from a valve is not sufficient to meet the supervision requirement.

12.6.4 Tamper Switches versus Initiating Circuits

Water flow devices that are alarm-initiating devices cannot be connected on the same initiating circuit as valve supervisory devices. This is commonly done in violation of the code by connecting the valve tamper switch in series with the initiating circuit's EOL device, resulting in a trouble signal when activated. This method of wiring does not provide for a distinctive visual or audible signal. This statement is true for most conventional systems. It should be noted that at least one known addressable system has a listed module capable of distinctly separating the two types of signals. These devices must be wired on the same circuit to meet the standard.

> **NOTE**
> Addressable devices are not connected to initiating circuits. They are connected to signaling line circuits.

Fire Pumps

A fire pump system is necessary when the available water supply is not adequate, in pressure or volume, to supply the fire suppression water needs of a sprinkler or standpipe system. The fire pump system is comprised of several components, including the pump, a driver (typically electric or diesel), and a controller. A typical fire pump system is shown here.

26405-14_SA12.EPS

12.6.5 Supervisory versus Trouble Signals

A supervisory signal must be visually and audibly distinctive from both alarm signals and trouble signals. It must be possible to tell the difference between a fire alarm signal, a valve being off-normal (closed), and an open (broken wire) in the circuit.

12.6.6 Suppression Systems in High-Rise Buildings

Where a high-rise building is protected throughout by an approved, supervised automatic sprinkler system, valve supervision and water flow devices must be provided on each floor (*NFPA 101®*). In such buildings, a fire command center (central control system) is also required (*NFPA 101®*). Fire-resistive cable systems may be required (see *NEC Article 728*).

13.0.0 TROUBLESHOOTING

The troubleshooting approach to any fire alarm system is basically the same. Regardless of the situation or equipment, some basic steps can be followed to isolate problems:

Step 1 Know the equipment. For easy reference, keep specification sheets and instructions for commonly used and serviced equipment. Become familiar with the features of the equipment.

Step 2 Determine the symptoms. Try to make the system perform or fail to perform as it did when the problem was discovered.

Step 3 List possible causes. Write down everything that could possibly have caused the problem.

Step 4 Check the system systematically. Plan activities so that problem areas are not overlooked, in order to eliminate wasted time.

Step 5 Correct the problem. Once the problem has been located, repair it. If it is a component that cannot be repaired easily, replace the component.

Step 6 Test the system. After the initial problem has been corrected, thoroughly check all the functions and features of the system to make sure other problems are not present that were masked by the initial problem.

13.1.0 Alarm System Troubleshooting Guidelines

Figure 71 is a system troubleshooting chart, *Figure 72* is an alarm output troubleshooting chart, and *Figure 73* is an auxiliary power troubleshooting chart.

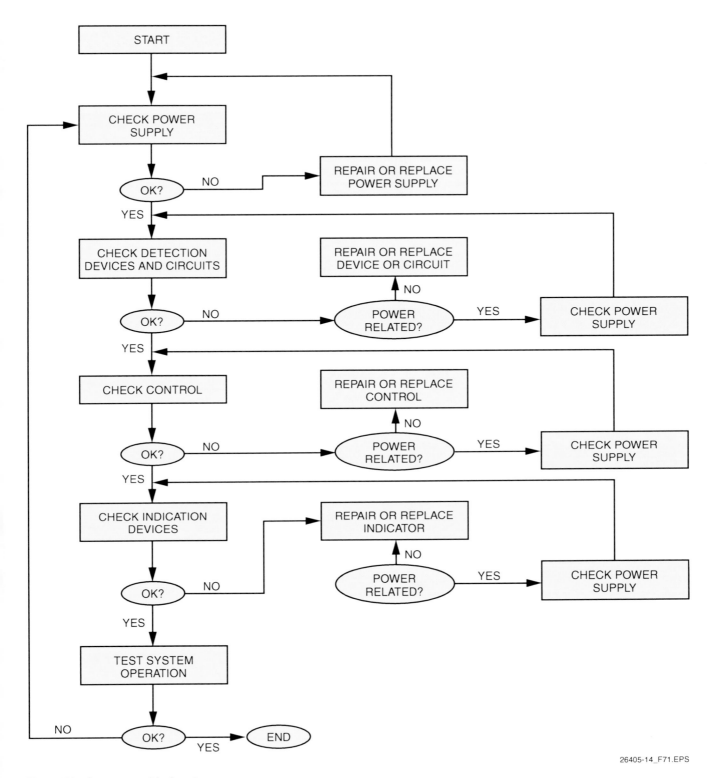

Figure 71 System troubleshooting.

26405-14_F71.EPS

The following guidelines provide information for resolving potential problems for specific conditions:

- *Sensors* – Always check sensor power, connections, environment, and settings. Lack of detection can be caused by a loose connection, obstacles in the area of the sensor, or a faulty unit. If unwanted alarms occur, recheck the installation for changing environmental factors. If none are present, cover or seal the sensor to confirm that the alarm originated with the

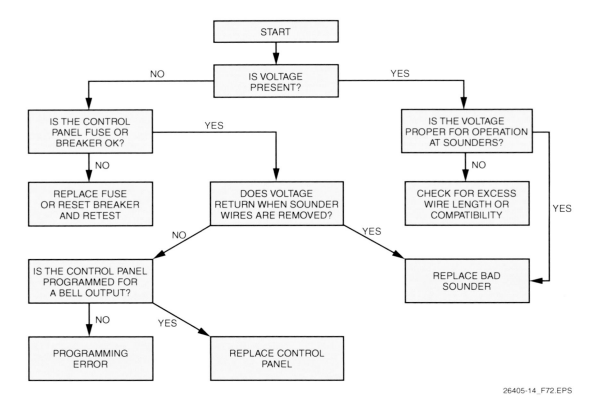

26405-14_F72.EPS

Figure 72 Alarm output troubleshooting.

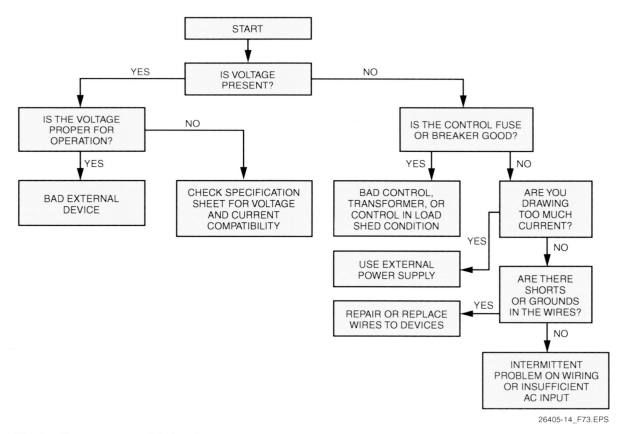

26405-14_F73.EPS

Figure 73 Auxiliary power troubleshooting.

sensor. If alarms still occur, wiring problems, power problems or electromagnetic interference (EMI) could be the cause. If the alarm stops when the sensor is covered or sealed, the environment monitored by the sensor is the source of the problem. Replacing the unit should be the last resort after performing the above checks.

- *Open circuit problems* – Opens will cause trouble signals and account for the largest percentage of faults. Major causes of open circuits are loose connections, wire breaks (ripped or cut), staple cuts, bad splices, cold-solder joints, wire fatigue due to flexing, and defective wire.

- *Short circuit or ground fault problems* – Shorts or ground faults on a circuit will cause alarms or trouble signals. Events occurring past a short will not be seen by the system. Some of the common causes of shorts or ground faults are staple cuts, sharp edge cuts, improper splices, moisture, and cold-flow of wiring insulation.

13.2.0 Addressable System Troubleshooting Guidelines

There is a wide variety of designs, equipment, and configurations of addressable fire alarm systems. For this reason, some manufacturers of addressable systems provide troubleshooting data in their service literature, but other manufacturers do not. Most of the general guidelines for troubleshooting discussed earlier also apply to troubleshooting addressable systems. In the event general troubleshooting methods fail to find the problem and no troubleshooting information is provided by the manufacturer in the service literature for the equipment, the best thing to do is to contact the manufacturer and ask for technical assistance. This will prevent a needless loss of time.

SUMMARY

A fire alarm system is a combination of components designed to detect and report fires, primarily for life safety purposes. A properly installed and fully functioning commercial or residential fire alarm system can save lives. This module covered the basics of fire alarm systems, components, and installation. Additionally, a basic introduction to some fire alarm system design criteria was provided.

A fire alarm system will not function properly if it isn't designed, installed, and maintained according to specifications. No two system installations will be exactly alike. To help regulate the different applications that exist, codes and standards have been established that dictate the specifics of design and installation. The electrician must be familiar with requirements of these codes and standards as well as the basic design, components, installation, testing, and maintenance of typical fire alarm systems.

1. The specific requirements for wiring and wiring equipment installation for fire protective signaling systems are covered in _____.
   a. *NFPA 70®*
   b. *NFPA 72®*
   c. *NFPA 101®*
   d. *NFPA 1*

2. A code that was established to help fire authorities continually develop safeguards against fire hazards is _____.
   a. *NFPA 1*
   b. *NFPA 72®*
   c. *NFPA 70®*
   d. *NFPA 101®*

3. A defined area within the boundaries of a fire alarm system is known as a(n) _____.
   a. section
   b. signal destination
   c. alarm unit area
   d. zone

4. A fire alarm system with zones that allow multiple signals from several sources to be sent and received over a single communication line is known as a(n) _____.
   a. addressable system
   b. conventional hardwired system
   c. zoned system
   d. multiplex system

5. An SLC circuit that consists of a Class A main trunk with a Class B spur circuit is known as an SLC _____.
   a. dual circuit
   b. combined circuit
   c. twin circuit
   d. hybrid circuit

6. An automatic detector that draws air from the protected area back to the detector is called a(n) _____.
   a. line detector
   b. spot detector
   c. air sampling detector
   d. addressable detector

7. A heat detector that is *not* a fixed temperature device is a _____.
   a. fusible link detector
   b. rate-of-rise detector
   c. quick metal detector
   d. bimetallic detector

8. Older type duct detectors used the _____.
   a. ionization principle
   b. light scattering principle
   c. light obscuration principle
   d. smoke rate compensation principle

9. The type of pull station that is restricted for use in special applications is the _____.
   a. single-action type
   b. glass-break type
   c. double-action type
   d. key-operated type

10. Addressable systems commonly use a verification method called _____.
    a. positive alarm sequence
    b. wait and check
    c. reset and resample
    d. cross-zoning

11. Fire warning system units for residential use must be listed in accordance with *UL Standard* _____.
    a. *780*
    b. *864*
    c. *985*
    d. *995*

12. Primary power circuits supplying fire alarm systems must be protected by circuit breakers not larger than _____.
    a. 15A
    b. 20A
    c. 25A
    d. 30A

13. The *NFPA 72®*–specified temporal three sound pattern for a fire alarm signal is _____.
    a. three short (½ sec.), three long (1 sec.), and three short, a pause, then a repeat
    b. three long (1 sec.), a pause, then a repeat
    c. three short (½ sec.), a pause, then a repeat
    d. three short (½ sec.), three long (1 sec.), three short, three long, a pause, then a repeat

14. Public area audible notification devices must have a minimum dB rating of _____.

    a. 65 at 10'
    b. 75 at 10'
    c. 85 at 10'
    d. 95 at 10'

15. The use of an attendant signal requires the approval of the AHJ.

    a. True
    b. False

16. Commercial fire alarm cables that run from floor to floor must be rated at least as _____.

    a. general-purpose cable
    b. water-resistant cable
    c. riser cable
    d. plenum cable

17. Power-limited fire alarm circuit conductors must be separated from nonpower-limited fire alarm circuit conductors by at least _____.

    a. 2"
    b. 6"
    c. 12"
    d. 24"

18. For smoke chamber definition purposes, an archway that extends down 18" or more from the ceiling is considered the same as a(n) _____.

    a. barrier
    b. boundary
    c. open grid
    d. beam pocket

19. The primary function of a commercial duct smoke detector is to _____.

    a. sound an alarm for duct smoke
    b. shut down an associated HVAC system
    c. detect external open-area fires
    d. disable a music system for voice evacuation purposes

20. In a supervised automatic sprinkler system, a visual or audible supervisory signal must be _____.

    a. different from an alarm signal
    b. the same as an alarm signal
    c. different from a trouble signal
    d. different from both the alarm and trouble signals

Module 26405-14
Supplemental Exercises

1. Provide the names of the NFPA codes listed below.

   NFPA 1 _____

   NFPA 70® _____

   NFPA 72® _____

   NFPA 101® _____

2. A(n) _____ fire alarm system allows multiple signals from several sources to be sent and received simultaneously over a single communication wire.

3. True or False? Smoke detection devices in an analog addressable fire alarm system make the decision internally regarding their alarm state.

4. A(n) _____ detector provides the best chance of detecting a fire during the incipient stage.
   a. photoelectric
   b. rate of rise
   c. ionization
   d. UV flame

5. What is generally used as the actuating medium in a rate-of-rise detector?

   _____

6. Name the two basic types of smoke detectors.

   _____

7. The melted and fused cable section in a(n) _____ detector must be replaced once it has been activated.

8. True or False? Primary power for a fire alarm system may be connected on the line side of the electrical main service disconnect switch.

9. All of the following devices receive code-authorized signals from the fire alarm control panel *except*

   _____.
   a. strobe devices
   b. alarm bells
   c. smoke detectors
   d. loud speakers

10. A digital alarm communicator _____ accepts and displays signals from a digital alarm communicator _____ sent over a public switched telephone network.

11. True or False? Cables installed in plenums and other air handling spaces may be secured using standard cable tie wraps.

12. A manual fire alarm box (pull box) installed at an exit door should be installed near the

    _____.

13. With regard to the area of coverage for detectors (the smoke chamber), all of the following are con-sidered a barrier *except* _____.
    a. an archway or doorway that extends more than 18″ down from the ceiling
    b. an open grid above a door or wall
    c. a space that is less than 18″ between the top of a low wall and the ceiling
    d. a space that is less than 12″ between the top of a low wall and the ceiling

14. True or False? Under normal conditions, smoke detectors should be installed after construction cleanup is completed.

15. Sidewall-mounted audible notification devices must be mounted at least _____ inches above the finished floor level (AFF) or at least _____ inches below the finished ceiling.

Trade Terms Introduced in This Module

Addressable device: A fire alarm system component with discrete identification that can have its status individually identified or that is used to individually control other functions.

Air sampling detector: A detector consisting of piping or tubing distribution from the detector unit to the area or areas to be protected. An air pump draws air from the protected area back to the detector through the air sampling ports and piping or tubing. At the detector, the air is analyzed for fire products.

Alarm: In fire systems, a warning of fire danger.

Alarm signal: A signal indicating an emergency requiring immediate action, such as an alarm for fire from a manual station, water flow device, or automatic fire alarm system.

Alarm verification: A feature of a fire control panel that allows for a delay in the activation of alarms upon receiving an initiating signal from one of its circuits. Alarm verification must not be longer than three minutes, but can be adjustable from 0 to 3 minutes to allow supervising personnel to check the alarm. Alarm verification is commonly used in hotels, motels, hospitals, and institutions with large numbers of smoke detectors.

Americans with Disabilities Act (ADA): An act of Congress intended to ensure civil rights for physically challenged people.

Approved: Acceptable to the authority having jurisdiction.

Audible signal: An audible signal is the sound made by one or more audible indicating appliances, such as bells, chimes, horns, or speakers, in response to the operation of an initiating device.

Authority having jurisdiction (AHJ): The authority having jurisdiction is the organization, office, or individual responsible for approving equipment, installations, or procedures in a particular locality.

> **NOTE**
>
> The NFPA does not approve, inspect, or certify any installations, procedures, equipment, or materials, nor does it approve or evaluate testing laboratories. In determining the acceptability of installations, procedures, equipment, or materials, the authority having jurisdiction may base acceptance on compliance with NFPA, or other appropriate standards. In the absence of such standards, said authority may require evidence of proper installation, procedure, or use. The authority having jurisdiction may also refer to the listings or labeling practices of an organization concerned with product evaluations that is in a position to determine compliance with appropriate standards for the current production of listed items.

Automatic fire alarm system: A system in which all or some of the circuits are actuated by automatic devices, such as fire detectors, smoke detectors, heat detectors, and flame detectors.

CABO: Council of American Building Officials.

Ceiling: The upper surface of a space, regardless of height. Areas with a suspended ceiling would have two ceilings: one visible from the floor, and one above the suspended ceiling.

Ceiling height: The height from the continuous floor of a room to the continuous ceiling of a room or space.

Ceiling surface: Ceiling surfaces referred to in conjunction with the locations of initiating devices are as follows:

- *Beam construction* – Ceilings having solid structural or solid nonstructural members projecting down from the ceiling surface more than 4" (100mm) and spaced more than 3" (0.9m) center to center.

- *Girders* – Girders support beams or joists and run at right angles to the beams or joists. When the tops of girders are within 4" (100mm) of the ceiling, they are a factor in determining the number of detectors and are to be considered as beams. When the top of the girder is more than 4" (100mm) from the ceiling, it is not a factor in detector location.

Certification: A systematic program using randomly selected follow-up inspections of the certified system installed under the program, which allows the listing organization to verify that a fire alarm system complies with all the requirements of the *NFPA 72®* code. A system installed under such a program is identified by the issuance of a certificate and is designated as a certificated system.

Chimes: A single-stroke or vibrating audible signal appliance that has a xylophone-type striking bar.

Circuit: The conductors or radio channel as well as the associated equipment used to perform a definite function in connection with an alarm system.

Class A circuit: Class A refers to an arrangement of supervised initiating devices, signaling line circuits, or indicating appliance circuits (IAC) that prevents a single open or ground on the installation wiring of these circuits from causing loss of the system's intended function. It is also commonly known as a four-wire circuit.

Class B circuit: Class B refers to an arrangement of initiating devices, signaling lines, or indicating appliance circuits that does not prevent a single open or ground on the installation wiring of these circuits from causing loss of the system's intended function. It is commonly known as a two-wire circuit.

Coded signal: A signal pulsed in a prescribed code for each round of transmission. A minimum of three rounds and a minimum of three impulses are required for an alarm signal. A coded signal is usually used in a manually operated device on which the act of pulling a lever causes the transmission of not less than three rounds of coded alarm signals. These devices are similar to the non-coded type, except that instead of a manually operated switch, a mechanism to rotate a code wheel is utilized. Rotation of the code wheel, in turn, causes an electrical circuit to be alternately opened and closed, or closed and opened, thus sounding a coded alarm that identifies the location of the box.

The code wheel is cut for the individual code to be transmitted by the device and can operate by clockwork or an electric motor. Clockwork transmitters can be prewound or can be wound by the pulling of the alarm lever. Usually, the box is designed to repeat its code four times before automatically coming to rest. Prewound transmitters must sound a trouble signal when they require rewinding. Solid-state electronic coding devices are also used in conjunction with the fire alarm control panel to produce coded sounding of the system's audible signaling appliances.

Control unit: A device with the control circuits necessary to furnish power to a fire alarm system, receive signals from alarm initiating devices (and transmit them to audible alarm indicating appliances and accessory equipment), and electrically supervise the system installation wiring and primary (main) power. The control unit can be contained in one or more cabinets in adjacent or remote locations.

Digital Alarm Communicator Receiver (DACR): A system component that will accept and display signals from digital alarm communicator transmitters (DACTs) sent over public switched telephone networks.

Digital Alarm Communicator System (DACS): A system in which signals are transmitted from a digital alarm communicator transmitter (DACT) located at the protected premises through the public switched telephone network to a digital alarm communicator receiver (DACR).

Digital Alarm Communicator Transmitter (DACT): A system component at the protected premises to which initiating devices or groups of devices are connected. The DACT will seize the connected telephone line, dial a preselected number to connect to a DACR, and transmit signals indicating a status change of the initiating device.

End-of-line (EOL) device: A device used to terminate a supervised circuit. An EOL is normally a resistor or a diode placed at the end of a two-wire circuit to maintain supervision.

Fault: An open, ground, or short condition on any line(s) extending from a control unit, which could prevent normal operation.

Fire: A chemical reaction between oxygen and a combustible material where rapid oxidation may cause the release of heat, light, flame, and smoke.

Flame detector: A device that detects the infrared, ultraviolet, or visible radiation produced by a fire. Some devices are also capable of detecting the flicker rate (frequency) of the flame.

General alarm: A term usually applied to the simultaneous operation of all the audible alarm signals on a system, to indicate the need for evacuation of a building.

Ground fault: A condition in which the resistance between a conductor and ground reaches an unacceptably low level.

Heat detector: A device that detects abnormally high temperature or rate-of-temperature rise.

Horn: An audible signal appliance in which energy produces a sound by imparting motion to a flexible component that vibrates at some nominal frequency.

Indicating device: Any audible or visible signal employed to indicate a fire, supervisory, or trouble condition. Examples of audible signal appliances are bells, horns, sirens, electronic horns, buzzers, and chimes. A visible indicator consists of an incandescent lamp, strobe lamp, mechanical target or flag, meter deflection, or the equivalent. Also called a notification device (appliance).

Initiating device: A manually or automatically operated device, the normal intended operation of which results in a fire alarm or supervisory signal indication from the control unit. Examples of alarm signal initiating devices are thermostats, manual boxes (stations), smoke detectors, and water flow devices. Examples of supervisory signal initiating devices are water level indicators, sprinkler system valve-position switches, pressure supervisory switches, and water temperature switches.

Initiating device circuit (IDC): A circuit to which automatic or manual signal-initiating devices such as fire alarm manual boxes (pull stations), heat and smoke detectors, and water flow alarm devices are connected.

Labeled: Equipment or materials to which has been attached a label, symbol, or other identifying mark of an organization acceptable to the authority having jurisdiction, which is concerned with product evaluation and whose listing states either that the equipment or material meets appropriate standards or has been tested and found suitable for use in a specified manner.

Light scattering: The action of light being reflected or refracted off particles of combustion, for detection in a modern day photoelectric smoke detector. This is called the Tyndall effect.

Listed: Equipment or materials included in a list published by an organization acceptable to the authority having jurisdiction that is concerned with product evaluation and whose listing states either that the equipment or materials meets appropriate standards or has been tested and found suitable for use in a specified manner.

> **NOTE**
>
> The means for identifying listed equipment may vary for each organization concerned with product evaluation [UL (Underwriters Laboratories), FM (Factory Mutual), IRI (Industrial Risk Insurers), and so on]. Some organizations do not recognize equipment listed unless it is also labeled. The authority having jurisdiction should use the system employed by the listed organization to identify a listed product.

Maintenance: Repair service, including periodic inspections and tests, required to keep the protective signaling system and its component parts in an operative condition at all times. This is used in conjunction with replacement of the system and its components when for any reason they become undependable or inoperative.

Multiplexing: A signaling method that uses wire path, cable carrier, radio, fiber optics, or a combination of these techniques, and characterized by the simultaneous or sequential (or both simultaneous and sequential) transmission and reception of multiple signals in a communication channel including means of positively identifying each signal.

National Fire Alarm Code®: This is the update of the NFPA standards book that contains the former *NFPA 71*, *NFPA 72®*, and *NFPA 74* standards, as well as the *NFPA 1221* standard. The NFAC was adopted and became effective May 1993.

National Fire Protection Association (NFPA): The NFPA administers the development and publishing of codes, standards, and other materials concerning all phases of fire safety.

Noise: A term used in electronics to cover all types of unwanted electrical signals. Noise signals originate from numerous sources, such as fluorescent lamps, walkie-talkies, amateur and CB radios, machines being switched on and off, and power surges. Today, equipment must tolerate increasing amounts of electrical interference, and the quality of equipment depends on how much noise it can ignore and withstand.

Non-coded signal: A signal from any indicating appliance that is continuously energized.

Notification device (appliance): See *indicating device (appliance)*.

Obscuration: A reduction in the atmospheric transparency caused by smoke, usually expressed in percent per foot.

Path (pathway): Any conductor, optic fiber, radio carrier, or other means for transmitting fire alarm system information between two or more locations.

Photoelectric smoke detector: A detector employing the photoelectric principle of operation using either the obscuration effect or the light-scattering effect for detecting smoke in its chamber.

Positive alarm sequence: An automatic sequence that results in an alarm signal, even when manually delayed for investigation, unless the system is reset.

Power supply: A source of electrical operating power, including the circuits and terminations connecting it to the dependent system components.

Projected beam smoke detector: A type of photoelectric light-obscuration smoke detector in which the beam spans the protected area.

Protected premises: The physical location protected by a fire alarm system.

Protected premises (local) fire alarm system: A protected premises system that sounds an alarm at the protected premises as the result of the manual operation of a fire alarm box, or as a result of the operation of protection equipment or systems, such as water flowing in a sprinkler system, the discharge of carbon dioxide, the detection of smoke, or the detection of heat.

Public Switched Telephone Network: An assembly of communications facilities and central office equipment, operated jointly by authorized common carriers, that provides the general public with the ability to establish communications channels via discrete dialing codes.

Rate compensation detector: A device that responds when the temperature of the air surrounding the device reaches a predetermined level, regardless of the rate-of-temperature rise.

Rate-of-rise detector: A device that responds when the temperature rises at a rate exceeding a predetermined value.

Remote supervising station fire alarm system: A system installed in accordance with the applicable code to transmit alarm, supervisory, and trouble signals from one or more protected premises to a remote location where appropriate action is taken.

Reset: A control function that attempts to return a system or device to its normal, non-alarm state.

Signal: A status indication communicated by electrical or other means.

Signaling line circuits (SLCs): A circuit or path between any combination of circuit interfaces, control units, or transmitters over which multiple system input signals or output signals (or both input signals and output signals) are carried.

Smoke detector: A device that detects visible or invisible particles of combustion.

Spacing: A horizontally measured dimension related to the allowable coverage of fire detectors.

Spot-type detector: A device in which the detecting element is concentrated at a particular location. Typical examples are bimetallic detectors, fusible alloy detectors, certain pneumatic rate-of-rise detectors, certain smoke detectors, and thermoelectric detectors.

Stratification: The phenomenon in which the upward movement of smoke and gases ceases due to a loss of buoyancy.

Supervisory signal: A signal indicating the need for action in connection with the supervision of guard tours, the fire suppression systems or equipment, or the maintenance features of related systems.

System unit: The active subassemblies at the central station used for signal receiving, processing, display, or recording of status change signals. The failure of one of these subassemblies causes the loss of a number of alarm signals by that unit.

Transmitter: A system component that provides an interface between the transmission channel and signaling line circuits, initiating device circuits, or control units.

Trouble signal: A signal initiated by the fire alarm system or device that indicates a fault in a monitored circuit or component.

Visible notification appliance: A notification appliance that alerts by the sense of sight.

Wavelength: The distance between peaks of a sinusoidal wave. All radiant energy can be described as a wave having a wavelength. Wavelength serves as the unit of measure for distinguishing between different parts of the spectrum. Wavelengths are measured in microns, nanometers, or angstroms.

Wide Area Telephone Service (WATS): Telephone company service that provides reduced costs for certain telephone call arrangements. In-WATS or 800-number service calls can be placed from anywhere in the continental United States to the called party at no cost to the calling party. Out-WATS is a service whereby, for a flat-rate charge, dependent on the total duration of all such calls, a subscriber can make an unlimited number of calls within a prescribed area from a particular telephone terminal without the registration of individual call charges.

Zone: A defined area within the protected premises. A zone can define an area from which a signal can be received, an area to which a signal can be sent, or an area in which a form of control can be executed.

Additional Resources

This module is intended to present thorough resources for task training. The following reference works are suggested for further study.

Certified Alarm Technician Level 1, Latest Edition. Silver Spring, MD: National Burglar and Fire Alarm Association.

Practical Fire Alarm Course, Latest Edition. Silver Spring, MD: National Burglar and Fire Alarm Association.

Understanding Alarm Systems, Latest Edition. Silver Spring, MD: National Burglar and Fire Alarm Association.

Figure Credits

NCCER CURRICULA — USER UPDATE

NCCER makes every effort to keep its textbooks up-to-date and free of technical errors. We appreciate your help in this process. If you find an error, a typographical mistake, or an inaccuracy in NCCER's curricula, please fill out this form (or a photocopy), or complete the online form at **www.nccer.org/olf**. Be sure to include the exact module ID number, page number, a detailed description, and your recommended correction. Your input will be brought to the attention of the Authoring Team. Thank you for your assistance.

Instructors – If you have an idea for improving this textbook, or have found that additional materials were necessary to teach this module effectively, please let us know so that we may present your suggestions to the Authoring Team.

NCCER Product Development and Revision
13614 Progress Blvd., Alachua, FL 32615

Email: curriculum@nccer.org
Online: www.nccer.org/olf

❏ Trainee Guide   ❏ AIG   ❏ Exam   ❏ PowerPoints   Other _____

Craft / Level: _____   Copyright Date: _____

Module ID Number / Title: _____

Section Number(s): _____

Description: _____

_____

_____

_____

_____

Recommended Correction: _____

_____

_____

_____

_____

Your Name: _____

Address: _____

_____

Email: _____   Phone: _____